The Dream
of Spaceflight

The Dream
of Spaceflight

Essays on the
Near Edge of Infinity

Wyn Wachhorst

BASIC
BOOKS

A Member of
the Perseus Books Group

Copyright 2000 by Wyn Wachhorst

Published by Basic Books,
A Member of the Perseus Books Group

Designed by Rachel Hegarty

Library of Congress Cataloging-in-Publication Data
Wachhorst, Wyn
The dream of spaceflight / Wyn Wachhorst.
    p.   cm.
  Includes bibliographical references and index.
  ISBN 0-465-09057-5
  1. Manned space flight.   I. Title.
TL873.W33 2000
629.45—dc21   99-086067
                                                       CIP

00 01 02 03 / 10   9   8   7   6   5   4   3   2   1

In memory of my father
Newton Edwin Wachhorst
abiding like the steeple in the village
and of my mother
Norma Harvey Wachhorst
whose passing predestined this book

The mind of the scientist, exploring space and matter, is closely related to the mind of the poet, whose task is to explore inner space and the reality of things. Like the scientist the poet is enchanted with an expanding universe of knowledge; but he keeps insisting that the new data must be incorporated into a moral universe, the universe that poetry originally created as myth and for which he must perpetually seek new metaphors.

Stanley Kunitz
A Kind of Order, a Kind of Folly

Contents

Acknowledgments   xi
Foreword by Buzz Aldrin   xiii
Preface: The Inner Reaches of Outer Space   xvii

1 Kepler's Children:
   The Dream of Spaceflight   1

2 The Romance of Spaceflight:
   Nostalgia for a Bygone Future   43

3 Seeking the Center at the Edge   83

4 Abandon in Place   119

5 Reflexions   159

Appendix: Space Chronology   173
Notes   179
Bibliography   197
Index   215

Acknowledgments

I am indebted to Buzz Aldrin, whose support has been indispensable; also to Dr. Lee Enright, Staige Blackford, J. D. McClatchy, Heidi Downey, Tom Jenks, Michelle Gillett, my agents Michael Larsen and Elizabeth Pomada, and my editor at Basic Books, William Frucht. Nor can I forgo mentioning two teachers who tower over all others in my 28 years of schooling—Crayton Thorup and Ted Hinckley. But above all, my wife, Rita, the rainbow through the gray mists of a writing life.

Foreword

Space journalists have long lamented the lack of poet-astronauts, hoping, perhaps, that pilots, engineers, and scientists might capture what has eluded the writers themselves. A half-century of space writing has seemed to harbor a tacit belief that the deeper motives for spaceflight are inexpressible. There are, of course, valuable histories and an array of exciting projections, but beyond a few well-worn phrases, space writers have implied by omission that the inner meaning is ineffable.

Wyn Wachhorst's book is thus a long-awaited departure. Apparently the inner experience *is* expressible. Poised between poetry and psychology, philosophy and history, the book is written in a lyrical style reminiscent of Loren Eiseley, Lewis Thomas, or Chet Raymo's *The Soul of the Night*. Probing beneath the political, economic, and tribal rationales, Wachhorst reminds us that the cold war alone cannot account for Apollo.

Watching the moonwalks on film, we feel like it all happened yesterday, so titanic were the achievements of Apollo and so timid our subsequent efforts. Yet those images are now more than a quarter-century old. Robotic probes have returned impressive pictures and invaluable information, but if you send a robot with a camera to Paris and peruse the pictures at home, you haven't really done Paris. As this book contends, it is humans who must go into space, to "wander far worlds and meet once more the dread unknowns, the dry-mouthed fears of the old explorers." The people who settled our continent were not afraid of risk; and beyond personal ambition, there was also a desire to be part of something larger, something epochal. If we balk before the challenge of space we will become less than the people who lifted us into the present.

Of course our ancestors did not disembark at Jamestown and Plymouth and suddenly build Los Angeles. A lasting human presence in space won't result from sudden leaps like Apollo; it has to move outward on a broad base of permanent support. But what we lack at present is less the technology than the vision. Beyond robotics and Earth-serving space stations lies the infinite journey. We have covered the globe in this millennium, and we will inhabit the solar system in the next. Escaping dependence on one vulnerable world, we will found new cultures and new species of awareness, spreading consciousness into the cosmos.

This is a truly outstanding book, intellectually vibrant, thematically ambitious, acutely independent-minded, and downright moving. Weaving history, poetry, and personal narrative, Wyn Wachhorst has written a book unlike any before, a merging of inner and outer that offers a new way of looking at spaceflight. My hope is that it will find a large audience.

The Dream of Spaceflight is not just for space buffs, it is for everyone who sees the mystery of the cosmos as analog to the human soul. It is for anyone who seeks a meaning beyond the self, a destiny beyond ephemeral ills. Probing the soul of exploration, it reaffirms the nobility of the human species, "an imperfect people of irrepressible spirit . . . who dare to dream of reaching the stars."

Many men go fishing their entire lives
without knowing it is not fish they are after.

Henry David Thoreau

The Inner Reaches of Outer Space

Perhaps it is more than coincidence that Sigmund Freud and Edwin Hubble shared the same moment in history, Freud exposing the rational mind as a tiny clearing in the dark forest of the soul, Hubble revealing that our galaxy is only one among billions, that the heavens are immense beyond imagination. To gaze into the night sky and feel the vastness and passion of creation is to glimpse an equally vast interior. We are aware of the stars only because we have evolved a corresponding inner space.

Beyond all the political and economic rationales, spaceflight is a spiritual quest in the broadest sense, one promising a revitalization of humanity and a rebirth of hope no less profound than the great opening out of mind and spirit at the dawn of the modern age.

The moon landing will be seen, a thousand years hence, as the signature of our century. It stands with the cathedrals and pyramids among those epic social feats that embody the spirit of an age. They are the dreams of the child in man, arising less from the ethic of work than from the spirit of play, rooted less in means than in meaning itself. Riding the crest of evolution, we explore our horizons as children probe their world in play, longing to perfect a grand internal model of reality, to find the center by completing the edge.

Living systems reach out to their environment, merging with larger systems in the fight against entropy. We know from the new science of chaos and complexity that an open system "perturbed" at its frontier may restructure itself, escaping into higher order. The quest for the larger reality is the basic imperative of consciousness, the hallmark of our species. Living systems cannot remain static; they evolve or decline. They explore or expire. The inner experience of this drive is curiosity and awe—the sense of wonder.

My purpose has been to bring the deeper, more personal concerns of philosophy to the subject of space exploration, to bond the wonder of the night sky to the workings of the human psyche. Probing the poetic perception of our first steps out into the cosmos, I have tried to capture the inner, subjective sense of what it means to explore other worlds, and to place that larger meaning in historical perspective.

Perhaps the deeper motifs are more suited to poetry or music—an "Apollo" symphony or a "Voyager" concerto. Much of the book is in fact a series of prose movements—a montage of images and reflections on the dream of spaceflight, from the romantic vision in the decades prior to *Sputnik*, when fantasy and reality seemed almost in balance, to the meaning of the moon landings and their premature demise. It is also a confluence of eclectic ideas, offering new perspectives on the psychology of wonder and the significance of the spacefaring vision in the evolution of Western culture. Perhaps the core message is that evolution and exploration are inseparable, that the curious intellect and the wondering spirit are the evolutionary process seen from inside. They are the hallmarks of humanity and the true propellants of spaceflight.

Though it may be millennia before spacefaring humanity views its past as either the dawn of a new world or an exile from Eden, the leap into space has profound psychological, philosophical, and spiritual implications for our own time. Yet among hundreds of nontechnical, nonfiction books in English on spaceflight, only three or four show more than a passing interest in its psychocultural significance. "The horror of the twentieth century," Norman Mailer once wrote, "was the size of each new event and the paucity of its reverberation."[1] It seems time, more than a quarter-century after the last moonwalk, to look for the larger meaning of humanity in space.

The Dream
of Spaceflight

Let us honor if we can
The vertical man
Though we value none
But the horizontal one.

W. H. Auden,
Epigraph for Poems

Chapter One

Kepler's Children: The Dream of Spaceflight

PULLED BY THE HAND, the six-year-old boy ran beside his mother, scurrying through narrow cobblestone streets. Peddlers pushed their iron-wheeled carts, players and musicians passed in procession, and lepers shook their rattles as night fell on the Swabian village of Weil. At the edge of town, the boy and his mother moved across a new-mown field and up a small hill in the last purple aura of dusk. The sounds of the village gave way to a silent canopy of stars, an ocean of light spanning the world from edge to edge, dwarfing the stern little steeple and its flock of houses, huddled in the twilight of an age.

The boy and his mother stood on the crest of the hill in the drone of the night wind. They had come to behold a streak of cosmic fire hanging motionless in the heavens, slashing across a full third of the sky. It was a moment burned into the boy's memory. Though his mother had heard that the comet foretold a new age, she would never know that there at her side stood the earthly counterpart to that cosmic aberration, who would pass through the world like the wind in the grass, sending a shudder through the whole medieval order, sowing the seed of the modern age. For the Great Comet of 1577 had fired the soul of Johannes Kepler.

He was a spindly, mange-eaten boy with a bloated face—sickly, pasty, chronically covered with scabs, boils, violent rashes, and putrid sores, plagued with worms, piles, myopia, and multiple vision. A precocious adolescent, neurotic, self-loathing, arrogant, and vociferous, he was an introspective loner, belligerently defensive, and prone to wild fits of anger. He was frequently beaten up for being an intolerable egghead, welcoming, perhaps, the periods of hard field labor that kept him from school. Raised by an extended family that Arthur Koestler describes as "mostly degenerates and psychopaths," the boy saw little of his vicious father, a mercenary adventurer who barely escaped the gallows, and who finally wandered away forever. Nor was he a special concern of his mother, a quarrelsome eccentric who was nearly burned as a witch.

Out of this "childhood in hell," as Koestler put it, came "the most reckless and erratic spiritual adventurer of the scientific revolution." His lifelong quest for larger meaning, for celestial harmony in an age of upheaval, sought a new image of God to transcend the misfortunes of a persecuted exile, swept about in the storms of seventeenth-century Europe. A strange and tormented genius whom Kant called "the most acute thinker ever born," Johannes Kepler stood astride the intellectual divide between medieval and modern.

The Poetic Structure of the World

Like the seafarers of their time, seventeenth-century scientists left well-worn paths to seek their own East in the west, for the new empiricism was often incidental to their search for a harmony and symmetry that would reveal the mind of God in nature. Kepler's laws of planetary motion—the first "natural laws" in the modern sense—not only rescued the Copernican system from philosophic obscurity but were also prerequisite to the law of gravity, upon which Newton built the modern universe. Yet the founding of modern science was a byproduct of Kepler's larger purpose, which was to seek no less than the poetic structure of the world, the grand geometric symmetry of all creation. His famous empirical laws were to him mere tools in a life-long obsession to prove his mistaken notion that the vertices of the five symmetrical solids (pyramid, cube, etc.), inscribed or nested one within the other, defined the orbital spheres of the six known planets, and that their relative motions matched the mathematics of musical harmony—the music of the spheres.

Driven by a cosmic vision, a grandiose fugue woven of science, poetry, philosophy, theology, and Pythagorean mysticism, Kepler's transcendent goal—to map the mind of God—was more medieval than modern. Yet in this quest for his summa of the Renaissance, the perfection of which eluded him all his life, Kepler had seen that empirical observation was key. He founded celestial mechanics, moving astronomy from theology to physics. Ironically, he compared his voyage of discovery to that of Columbus, little knowing that he himself had discovered his America believing it was India. It was neither Copernicus nor Galileo but Johannes Kepler who launched the scientific revolution.[1]

Failing to match the paths of the planets to the vertices of his symmetrical solids, Kepler came finally to question the orbits themselves. By a stroke of luck, the astronomer Tycho Brahe, whose observations were the most accurate to date, was so protective of his data that he would give Kepler only his figures for Mars. Only the orbit of Mars, having the largest eccentricity, could have forced Kepler to abandon the dream of symmetry. Attempting to derive the orbit from Brahe's data, he was unwilling to accept a discrepancy of even a few minutes of arc. He labored six years on the problem, covering nine hundred folio pages in small handwriting, at one point repeating seventy times a process involving thousands of calculations. Looking first for an egg-shaped orbit, he settled sadly on the lowly ellipse. Thus the planet Mars, so prominent in subsequent dreams of spaceflight, led Kepler to the first natural law: planets move in elliptical orbits. This in turn led him to two additional discoveries:

that a planet sweeps out equal areas in equal times, and that the squares of the periods of revolution of any two planets are as the cubes of their mean distances from the sun—the revelation that led Newton to the law of gravity. Though Kepler salvaged his vision of harmony with his notion of musical intervals, he had opened the universe to infinity. No longer in the supernatural province of God, the planets now belonged to man. Yet the three laws had no apparent relationship before the invention of calculus and analytic geometry; thus, one of Newton's greater achievements was to spot the laws in Kepler's writings, hidden away, as Koestler notes, "like forget-me-nots in a tropical flower bed." With those laws Newton laid the foundation of the pyramid that would put man on the moon.

Seeking the Past in the Future

"Let us create vessels and sails adjusted to the heavenly ether," Kepler wrote to Galileo, "and there will be plenty of people unafraid of the empty wastes. In the meantime, we shall prepare for the brave sky-travelers maps of the celestial bodies."[2] The reduction of the planets from ethereal orbs to rocky worlds like our own resulted from both Kepler's work and the advent of the telescope (which was neither invented nor understood by Galileo; it was Kepler who deciphered its principles and founded the science of optics). The planets had become real places in the sky.

Thus Kepler's *Somnium* (1634), published four years after his death, became the first cosmic voyage in modern science fiction, providing a model for later works. Though the moonflight itself was a dream, Kepler took great pains to envision the lunar surface and its inhabitants. Disguised as fiction, it was in fact a heretical scientific hypothesis, the first seriously to suggest extraterrestrial life. Kepler poured his soul into *Somnium*. In it, observes Koestler, were "all the dragons which had beset his life—from the witch Fiolxhilda and her vanished husband, down to the poor reptilian creatures in perpetual flight, shedding their diseased skin."[3] And the same tension of old and new that had defined his life colored this last work. For if the hero was whisked to the moon by spirits, he was also first to see Earth afloat in the lunar sky, where the continent of Africa resembled a head, and Europe a girl bending down to kiss it.

In projecting a lunar reality based on the best science of the day, *Somnium* was ahead of its time. Though the literary history of voyages beyond the Earth dates from second-century Greece, tales before the nineteenth century were the tongue-in-cheek gimmicks of sermon and satire. Travelers were swept to the moon by angels and witches, by ships borne on whirlwinds or waterspouts, and in chariots drawn by flame-red horses. One voyager wore the wings of a vulture; others sailed an angular framework suspended from a team of swans, donned bottles of dew that rose with the dawn, or ascended the heavens in an iron car by lobbing lodestones continuously upward. Shrewdest of all, perhaps, were those who simply arrived on the moon unexplained. But even these tales were virtually nonexistent until the fabulous voyage flowered in Elizabethan literature.[4]

The sudden popularity of the cosmic voyage in the 1630s found fertile ground in an ambivalence toward accelerating change, a yearning to escape the present by seeking a simpler past in a purified future. Like twentieth-century America, early seventeenth-century England straddled old and new, oscillating between extremes of hope and pessimism, experiencing what literary historians have called a failure of nerve, a metaphysical shudder. Amid political and economic turmoil, the medieval world-picture was surfacing into critical consciousness. Like present-day materialism, it was losing the power to provide axiomatic order and meaning.[5] But if a melancholy sense of mutability and decay reinforced the Christian notion that history must unwind to its end, the new visions of Kepler, Galileo, Bacon, and Descartes, with their expansive belief in humanity's ability to improve on nature, promised escape from history. The New World became the idyllic setting for utopian tales, and Kepler's notion of a world in the moon prompted fables of escape to simpler but superior peoples on exemplary planets.

The vast resources of the New World so liberated human potential that the imaginative mind felt a new relation with the universe—a new sense of control over destiny. But the consequent rise of the self-reflexive individual, severed from institutional contexts of identity, brought the loss of innocence that has made the reconquest of Eden the organizing image of the last half-millennium.[6] If the Copernican expulsion from the literal center symbolizes this loss of larger meaning, then the leitmotif of the longing to return—from Kepler in the seventeenth century to von Braun in the twentieth—has been the dream of spaceflight.

In contrast to the horizontal expansion of modern materialism, which has confused Paradise with power itself, this longing for meaning has been a vertical vision—seeking the whole over the part, the why over the how, meaningful ends over endless means—akin to the medieval penchant for spirit over matter, which put final meaning at the highest point in the heavens, embodied in the cathedral's soaring vaults. From early explorers, scientists, and novelists to pulp writers and rocket pioneers, the forerunners of spaceflight have been keepers of this lost vertical vision. Imbued with curiosity and wonder, they have carried the quest for meaning across the modern desert, seeking the East in the West, the past in the future, the center at the edge.

Rites of Passage

Certainly curiosity has been prerequisite to our success as a species. But what drove a man like Magellan, who sought the mythic *el paso*, the passage to the Orient, only to find himself on the far side of a globe many times larger than imagined, hurled about in howling winds amid mountains of water, or becalmed in a stillness that roasted the flesh, putrefied meat, and fouled water casks with green scum? He and his men ate worm-filled biscuits turned to powder and the ox hides off the masts, unable to get their fill of that delicacy the rat. And what drove the Spanish adventurer who trekked across the Isthmus of Panama in 1513 with 190 men in heavy armor, scaling steep mountains, forging defiles and dark troughs of swampy rain forest, beset by snakes, predators, and poison arrows? A nineteenth-century expedition following the same route lost every man; even today no road traverses it. But the hardy redhead Spaniard reached the westernmost mountain with sixty-seven men. Climbing the last peak alone with his dog, Vasco Nuñez de Balboa stood silent on the bare summit. Beyond the virgin forest glistened the vast blue of the Pacific stretching away with the other half of the world.

The price of exploration has been high. On the far side of the Earth, Magellan's body, mutilated by native knives and spears, lay on a barren beach. Balboa, a victim of political intrigue, was beheaded in a public square, his head displayed on a stake, and his body thrown to the vultures. What motivates those who venture over the edge, who trek over barren plains, through tangled jungle, or across the Arctic waste; who ride on fire over the rocks of the moon, only to yearn for the relentless red desert of Mars? Surely it is more than gold and glory. What the biographies of most explorers reveal, in fact, is a sometimes selfless obsession with reaching the pristine edges of reality. At the heart of exploration, it seems, is the attempt to complete the grand internal model of reality, to broaden the context of meaning, to find the center by completing the edge.

In a sense, the age of exploration, geographic and scientific, is the whole of modern history, a five-century quest for self-definition. Just as the adolescent's desire to remain dependent in a maternal Eden conflicts with a growing sense of autonomy, the modern age, suspended between an authoritarian past and an existential future, has been man's rite of passage. The adolescence of humanity began with the age of exploration.

So it is fitting that the impetus for Kepler's grand internal model of the solar system was both a priori and inductive, ideal and real, regressive and progressive, a tension resembling that of adolescence. For not only was the seventeenth century a torn and restless era of nascent individualism, but the architects of the modern age seem themselves to have lived in that limbo between childhood and maturity. Copernicus found refuge from lonely isolation in secret and incessant elaboration of his system; Galileo was a self-centered egomaniac; and Newton, also a rational mystic, was a sickly recluse, alienated from his parents, obsessed with his work, and a lifelong celibate. Kepler himself, as Koestler observes, seemed polarized to a point verging on insanity. He was a "dark, wiry figure, charged with nervous energy," his Mephistophelian profile "belied by the melancholia of the soft, shortsighted eyes." He was both naïve and profound, fanatically patient and violently irritable, burdening his sweeping fantasies with pedantic obsessions over fraction-of-a-decimal deviations. Set against twenty years of dreary, heartbreaking computations, the true significance of which he never perceived, his labors were suggestive, says Koestler, of the "explosive yet painstakingly elaborate paintings by schizophrenics."[7] In that tension lay the spark of the scientific revolution and the dream of reaching the stars.

A Wind Between Worlds

Perhaps there was a brief moment, on some bare summit of imagination, when Kepler himself glimpsed the true scope of the cosmos, stretching away before him like Balboa's Pacific. But Kepler, too, had paid the price of beaching his boat on the shore of a new world. For attempting to stand above the political and religious forces of the Thirty Years War, he was excommunicated and exiled, forced to migrate from city to city with his family, his household goods piled in a wagon. He was chronically ill and ever in fear of penury and starvation, trekking from court to court in his baggy, food-stained suit, pleading for his fees. Seldom understood, he was desperately lonely amid the ignorance and provinciality of his time. His first wife, a nagging, simple-minded, sulking woman who viewed with contempt his position of stargazer, died of typhus at thirty-seven. In the end, Kepler lost six of his twelve children. His mother, accused of witchcraft and chained in prison for fourteen months, died shortly after he secured her release. Through all this, fluctuating between ecstatic discovery and frequent depression, Kepler wrote his *Harmony of the World*, convinced that the world that so mistreated him was nonetheless beautiful. One day in the fall of 1631 he set out on a skinny nag in search of funds with which to feed his children. Three days out, he died in a fever at the age of fifty-eight. As befit an outcast who wandered the edge, the tides of war erased his grave and all trace of his bones.

Like the Great Comet itself, the fugitive odyssey of Johannes Kepler drifted on that great divide between Aristotle and Newton, animism and mechanism, spirit and matter, a wind between worlds. For decades after his death, until Newton unearthed his significance, scholars saw only the wild-minded apriorist whose speculations had included an Earth soul and radiations from the planets that shaped human lives. The ironic postscript was that the planets, which his labors had so demystified, stood in the same positions at the moment of his death as they had on the day of his birth.

Embodying the vertical mind of the medieval while setting the modern mind in motion, the two sides of this wholly original man foretold the tension of our collective adolescence—the overt drive to recapture Eden with the rational mind and the covert need to restore the old mysteries, to humble man once more before Nature. Epitomizing this polarity, the prophets of spaceflight—from the astronomers who mapped imaginary Martian canals to the writers who peopled them with exotic beings to the rocket pioneers themselves—were realists in search of the ideal, reflecting the expansive spirit of the age while preserving the transcendent vision. A classic Heinlein story comes to mind in which the community aboard a starship, intended to reach its destination after countless generations, has forgotten the origin and purpose of the journey.[8] A similar loss of meaning has befallen spaceship Earth, where dreams of enlarging the grand internal model, of moving toward some final cosmic perspective, yield to the cancerous individualism that devours the social organism.

The Romance of the Red Planet

Two centuries after Kepler, when Enlightenment rationalism was itself becoming the exhausted old order, a cultural crisis similar to that of seventeenth-century England arose in the wake of the French Revolution amid the first signs of urban-industrial stress. Once again a literary awakening sought an Edenic past in a purified utopian future. As Enlightenment faith in the rational perfectibility of man descended from the moral to the material, the Romantic revolt seeded the future genres of popular culture. The list of nineteenth-century proto-science-fiction authors, from Mary Shelley to H. G. Wells, all of whom explored the dark side of the machine age, reads like a roll call of Romantic reaction.

But like its Romantic roots, science fiction has always reflected both progressive and regressive responses to social change. Its rising popularity over the past two centuries owes much, in fact, to its paradoxic embodiment of the ever more polarized modern temper. Exemplifying the nineteenth-century vision of technological salvation, the tales of Jules Verne were by far the most popular, looming large in the childhood of virtually every spaceflight pioneer. Hermann Oberth, the father of German rocketry, reread Verne's classic *From the Earth to the Moon* (1865) until he knew it by heart.[9]

But it was in America, Verne's land of tomorrow, that science fiction, the cosmic voyage, and the quest for an idealized past in a limitless future found their greatest appeal. A partial explanation may be that America has collectively reenacted the passage from childhood to adolescence. Settled by millennial sects and shaped by Enlightenment principles, America rode the leading edge of modern history. The absence of limits—freedom—became the core of the American faith, from the rejection of political and ecclesiastic authority to dreams of success, manifest destiny, and unlimited economic growth. Abundant resources, geographic insulation, waves of uprooted immigrants, a restless physical and social mobility, and the consequent lack of traditional institutions exacerbated the fragmenting forces of recent centuries, producing a pluralistic society of disconnected individuals, each free (and condemned) to find their own separate meanings.

Near the end of the nineteenth century, industrial expansion, a shrinking globe, the closing of Western civilization's four-century frontier, and the loss of local autonomy marked a transformation even more profound than that of the early seventeenth century. Just as in Elizabethan and Stuart England an organic medieval order gave way to a new existential freedom, late-nine-teenth-century Americans began the search for a mechanical order to counter the logical extremes of that freedom. The response resembled that of an adolescent encountering the world of communal responsibility: a wave of escapist fantasy swept over popular culture. An early sign was the dime novel, idealizing the cowboy as the last areas of free land disappeared. Soon, however, following the lead of Verne, Wells, Burroughs, Gernsback, Campbell, and others, science fiction became a popular alternative to the Western. It was fitting that this mythology appealed most strongly to adolescent males. Its writers were preponderantly outsiders, men who saw in science fiction a haven for the vertical vision, a means of transcending mundane modernity.[10]

A major catalyst was the rise of speculation on the possibility of life on Mars. It began with a Boston Brahmin and U.S. diplomat named Percival Lowell, brother of the poet Amy Lowell and Harvard president James Russell Lowell. Like Kepler, Lowell's earliest memory was of a comet—Donati's of 1858—which stretched across an entire quadrant of the sky. "I can see yet a small boy half way up a turning staircase," he wrote, "gazing with all his soul into the evening sky where the stranger stood."[11] Also like Kepler, his destiny was shaped by the planet Mars. Interpreting Schiaparelli's 1877 reports of Martian *canali* (channels) as literal canals, he devoted the later part of his life to a study of the planet, building an observatory on "Mars Hill" near Flagstaff, Arizona, and sighting more than seven hundred canals between 1894 and his death in 1916.

Over the years he developed the theory of an ancient, withered, Earthlike planet, a frigid desert irrigated by a vast system of canals built by an older, wiser civilization, heroically trying to delay extinction. Controversy raged for decades, with most astronomers denying any such canals, driving Lowell into long periods of nervous exhaustion. In the end, as the *Mariner* and *Viking* probes revealed, the canals were figments of a fervent and wishful imagination. Nothing on Lowell's maps matches the real topography of Mars, the features of which, in any event, would have been too small to see. Yet a few observers did see some of the same straight lines that Lowell had so carefully penciled on countless long cold nights at the twenty-six-inch telescope on Mars Hill. "I have the nagging suspicion," Carl Sagan has said, "that some essential feature of the Martian canal problem still remains undiscovered."[12]

Though what Lowell called his "far wandering" in the night sky went beyond Mars and laid groundwork for the discovery of Pluto, and for Edwin Hubble's expanding universe of numberless galaxies, it remains the larger legacy of this much-maligned man that the notion of inhabited planets and superior Martian intelligence passed into the public imagination, resuscitating the seventeenth-century belief that extraterrestrial life was commonplace and convincing young pioneers that exploring the planets was a human imperative. One of these was Robert Goddard, inventor of the liquid fuel rocket, who had been fascinated by Lowell's lectures at MIT. Between 1880 and World War I, the thirty-year Mars furor, which at its peak in 1907 overshadowed the national depression, engendered more than two hundred works in English on interplanetary voyages. So deeply had Lowell penetrated the public mind that the nation panicked when *The War of the Worlds*—the novel that won

H. G. Wells his initial fame—was dramatized as a simulated news broadcast in 1938. Mars had come to incarnate the mystery of the cosmos, a blood-red beacon burning low in the night sky, beckoning generations of youths to wander the grassy banks of the great canals, down long green valleys to the gates of crystal cities.

Another author who emerged from the Mars mania had no illusions of writing literature on a higher plane, yet he bears major responsibility for the mythic image of Mars. An ex-cavalry-man, gold prospector, railroad cop, and candy vendor, Edgar Rice Burroughs had gone from failure to failure, selling pencil sharpeners at the time he wrote *Under the Moons of Mars*, serialized in *All-Story Magazine* in 1912, the first of ten books about John Carter on Mars. With their clumsy prose, formulaic plots, and appeals to simple emotions, the books depict a dying world of moss-covered sea bottoms, lost cities, and exotic races. It is a world of decadent, sword-carrying barbarian cultures, living off superscientific technologies from an ancient past. In stories stuffed with Victorian values of love and honor, scantly clad women are threatened by hideous monsters and saved by muscular heroes who respect their virtue utterly. Yet Burroughs had a genuine inventiveness, a mastery at populating imaginary landscapes. A true sense of wonder surrounded the green, four-armed, fifteen-foot, sharp-tusked Tharks, the symbiotic relationship of the headless rykors and the bodiless kaldanes, and the beautiful, red-skinned, egg-laying princess Dejah Thoris.

Burroughs's stories of Barsoom, as he called Mars, launched the genre of scientific romance and space opera, capturing the likes of Carl Sagan, Arthur Clarke, and Ray Bradbury while spawning myriad progeny, from Buck Rogers and Flash Gordon to *Star Wars* and *Star Trek*. Bradbury notes that most of the scientists and astronauts he has met are beholden to some romantic encountered in childhood. "It is part of the nature of man," he adds, "to start with romance and build to reality. We need this thing which makes us sit bolt upright when we are nine or ten and say, 'I want to go out and devour the world.'"[13] Bradbury himself, who published his classic *Martian Chronicles* in 1950, owed a great debt to that pencil-sharpener salesman whose seventy-some books, with sales over 100 million, included adventures in the steamy jungles of Venus, hollow-

Earth stories, and tales of Tarzan. That same year, Edgar Rice Burroughs died at the age of seventy-four, at his home in Tarzana, California, sitting in bed, reading a comic book.

One-Dream Man

One of those whose imagination had been "gripped tremendously" by the *Boston Post's* serialization of *War of the Worlds* was a boy who went out to prune his grandmother's cherry tree one autumn afternoon in 1899. Lying in the high limbs and gazing into the New England sky, he suddenly envisioned a whirling machine, spinning until it began to lift, rising above his grandmother's house, his invalid mother, and his own chronic infirmity, out of Earth itself to the canals of Mars, and to the strange civilizations that beckoned from the pages of books he had read and reread. At a time when even the airplane was unknown, he conceived the dream that would consume his life: to find some principle, some mechanism, some great machine that would take man to the stars.

Robert Hutchings Goddard marked the day in his diary; he would remember it as a private anniversary during decades of disappointment, as the rockets he invented exploded, fizzled, or failed to ignite. Although the Russian Konstantin Tsiolkovsky had written on the potential of the rocket as early as 1903, Goddard independently came to the same conclusions after considering a number of methods, including multistage projectiles. Unlike Tsiolkovsky, however, Goddard was a practical engineer who went beyond theory to build and test working models, launching the world's first liquid fuel rocket in 1926.

Working with a handful of helpers in a place called Hell Pond, near Worcester, Massachusetts, and later in Eden Valley, in the desert of New Mexico, this stooped, tubercular man found meager financial support for his rockets until another soaring idealist, Charles Lindbergh, intervened on his behalf. A withdrawn professor of physics who wore shabby clothes for comfort, Goddard always concealed his cherry tree vision of Mars behind military and meteorological rationales. Meanwhile, military motives and the independent theories of Hermann Oberth, along with a standing order for all of Goddard's patents, put the Germans ahead. Yet the reclusive Goddard invented every principle of propulsion and guidance, anticipating all the essential elements of the modern multistage rockets that have taken man to the moon and sampled the soil of Mars.

From his first experiments on a farm in Auburn, Massachusetts, to his last days in the desert, through years of failure and "moonman" ridicule, this tireless man, who always laughed at the simplest of jokes and who was described by a friend as possessing an "overwhelming averageness," a "boy scout conventionality," remained the Yankee optimist. Now and then he returned to the cherry tree where his vision began, always noting the day in his diary. In it he once wrote, after his fiancée had tired of his obsession and left him: "God pity a one-dream man."[14]

In his later years, Goddard still labored from three in the morning until eight at night in the broiling desert heat, staggering home after each defeat with fatigue visible over his whole body, his shoulders bent forward to accommodate his weak lungs. "How many more years I shall be able to work on the problem I do not know," he once wrote to H. G. Wells. "There can be no thought of finishing, for 'aiming at the stars' is a problem to occupy generations. . . . No matter how much progress one makes, there is always the thrill of just beginning." When the spaceship was still pulp fantasy, he wrote of "morning in the desert, when the impossible not only seems possible, but easy."[15]

During his last days, before he succumbed at sixty-two to pulmonary problems and throat cancer, Goddard spent sleepless nights sitting outside in a blanket watching the stars. He died on August 10, 1945, the day after the atomic bomb ended the war and launched the space age. Only four years later, at White Sands, New Mexico, not far from Eden Valley, where Goddard's launch tower gathered rust, a small rocket, mounted on the V-2 that his work had inspired, crossed the threshold of space. What might Goddard, the "moony" and "crackpot," have felt had he seen the glowing arc of the Earth on the black of space—the vivid planet of swirling clouds, cerulean seas, and pastel continents, floating below orbiting astronauts like a great mothership?

Goddard's life belongs to the epic of humanity's great projects, from massive stone blocks creeping over the Egyptian desert to the gargantuan crawler inching Apollo toward the pad. Perhaps, as Esther Goddard sat in the VIP stands watching the launch of Apollo-Saturn to the moon, she recalled the predawn excursions into the desert with their rocket roped to a rickety truck. As Goddard had set up his crude apparatus against the first glow of day, Esther had readied the telemetry—the home movie camera, binoculars, an old alarm clock to drive a recording drum. Perhaps her memory of the home-made launch tower dissolved to the distant view of the great gantry at the cape as the thirty-six-story, three-thousand-eight-hundred-ton Saturn rocket rose slowly for a silent, shimmering moment—as though it were still a dream—then thundered into history.

The Pearl of Peenemünde

Also in the VIP stands for that launch was a little gray-haired man with sparkling eyes, a small moustache, and an eagle-beak nose. It is unlikely that many recognized him, despite his cameo appearances at major junctures in spaceflight history. A decade earlier at Redstone Arsenal in Alabama, where he was a token employee of the team that launched the first American satellite, he wandered about the buildings in similar anonymity, wearing a gray beret and carrying a briefcase that contained only his lunch. Even at Peenemünde on the Baltic, he was unfamiliar to many as he stood watching the first successful launch of the V-2. And in 1929, after he had nearly lost his sight and hearing in a disastrous attempt to build a working rocket for the premiere of the German film *Woman in the Moon*, he slipped unrecognized into a Berlin theater on the film's opening night and watched from a cheap seat. For he was a theorist, not an engineer; an absent-minded visionary, not a man of action, whose interests, like Kepler's, were philosophic at heart. Yet it was this man, more than anyone else, who set the space age in motion.

As a shy, gangling boy in Transylvania, Hermann Oberth had memorized Verne and dreamt of going to the moon, even practicing weightlessness in a swimming pool. To his father's disappointment, he abandoned the study of medicine for mathematics and physics, writing his thesis in 1923 on the rocket as the key to spaceflight. Finding no professor who would support the paper, he scrapped his scholastic hopes and published it on his own. A small, thin pamphlet ignored or belittled by the scientific establishment, *Die Rakete zu den Planetenraumen* (The Rocket into Planetary Space) sold out immediately and launched the dream of real spaceflight in the twentieth century. The book led directly to the founding of the German Rocket Society and inspired the eighteen-year-old Wernher von Braun to devote his life to reaching the moon. Von Braun and other members of the society were given $100 million by the German army to develop the V-2 at Peenemünde. It was this rocket team that later fled to America and ultimately put man on the moon. Had *Die Rakete* appeared later the V-2 would have missed its military moment, and the moon landing might have been delayed indefinitely.

Although Oberth had independently reached the same conclusions as Goddard, laying the theoretical groundwork for the V-2, he received little credit in the end. Von Braun and his team of engineers and military exploiters accomplished the dream while Oberth languished in the wings, ridiculed by scientists, strung along by financiers, and sidelined by governments. After the war, in which he had lost a daughter at Red Zipf test site and a son on the Russian front, he returned home to teach school for most of his remaining years. He had been the key innovator, making the initial sacrifices only to move directly from ridicule to obsolescence and pensioned obscurity. A few years before Oberth's death, Arthur Clarke spotted him among a crowd of visitors being guided through the great space center in Greenbelt, Maryland, that now bears Goddard's name. It was likely, noted Clarke, that even the young scientist guide had never heard of him.

In the end, no single figure emerges clearly as the father of spaceflight. Perhaps it was Goddard or Oberth or Tsiolkovsky or von Braun. Perhaps it was Sergei Korolev, the "Chief Designer," who survived Siberian prisons and lifelong anonymity to launch *Sputnik* and the space race. Although it is difficult to designate the man, the decisive moment is clear. Spaceflight was born on a brooding stretch of land along the Baltic Sea, at the mouth of the River Peene on the north coast of Germany. Hidden away in an ancient forest on the reedy marshlands of Peenemünde, a forty-seven-foot A4 rocket stood on the firing table like a monstrous mimicry of Wolgast Cathedral, visible on the western hills. Shedding its chain, the rocket rose out of the woods, howling to a record height of sixty miles on its first successful test. "Today," said team leader Walter Dornberger, "the spaceship was born." The day was October 3, 1942, and the fourteen-ton A4, poised like a Gothic spire against the tall trees, was later renamed *Vergeltungswaffe*, "vengeance weapon"—the V-2.

Everything about Peenemünde suggested the Gothic, from Hitler's own Wagnerian megalomania ("But what I want is annihilation, annihilating effect!") to the forest castle surrounded by missiles, where the team was decorated one night in a dark dining room, each presentation following a launch. A curtain was opened in the direction of each firing, allowing the room to flash with the light and reverberate to the thunder of the rocket. A tint of modern Gothic also touched Wernher von Braun, son of Baron Magnus von Braun and technical chief at Peenemünde. His fixation on spaceflight dated from his student days, when an experiment with a bicycle-wheel centrifuge had put a ring of mouse blood on the

walls of his rented room. At seventeen, he intro-
duced himself to the German Rocket Society—secretary Willy Ley came home one fall day in 1929
to find a polite young man playing the *Moonlight
Sonata* on his piano. Developing rockets for the
army during icy winter nights in the Brandenburg
woods south of Berlin, the cavalier von Braun
often lit them with a cigarette, graduating to a
twelve-foot match as the models grew larger, once
walking away unscathed when an explosion blew
the metal doors off the test shed, bent steel girders,
and embedded fragments in the trees. At Peenemünde, where he used the military to further his
covert vision of space, he would sit into the wee
hours of the morning sketching plans for flights to
the moon and Mars, reveling, as Dornberger rem-
inisced, in the dream of "anything that was big,
powerful, immeasurable and far in the future."[16]

But if Peenemünde was a labor of love for von
Braun and his teammates—who had danced,
laughed, leaped, and embraced as the V-2
soared away that crisp October day—it also ex-
acted a great price. For another twenty-three
months, solemn men in wave-beaten boats retrieved rocket parts from the bay in the icy dusk,
making another 65,000 changes before the V-2

was operational. In the process, 735 lives were lost in two bombing raids on Peenemünde. But the figure pales before the 8,380 V-2 casualties in London alone and the suffering of thousands of enslaved workers who assembled the rockets in the huge underground factory at Nordhausen. The vengeance weapon was the fruit of a vengeance war, similar in cost to the atomic bomb; together they would launch the dark side of the space race. Yet the V-2, conceived in suffering and sent in the wake of war to the deserts of the world, was reborn to its creators' purpose; its progeny will take us to the stars.

Vigil in the Desert

Some ninety miles north of El Paso, Texas, in the desolate Tularusa Basin of New Mexico, lay a barren expanse of flat desert called White Sands Proving Ground, a shimmering sea of sunbaked gypsum stretching to the horizon. Like black and white scenes from old science fiction films, a long, straight road followed power poles over the desert to distant mountains. Robert Goddard had spent his last tortuous years testing rockets in the neighboring county. A few miles away, the first atomic bomb had turned the terrain to glass. It was here that the V-2s were redeemed. Symbols of the postwar promise, they stood like steeples against the dawn sky—Goddard's "morning in the desert," when all things seemed possible. In 1950, when I was carving balsa copies of the silver ship from *Destination Moon*, inserting CO_2 capsules to send them hissing in cartwheels over the neighbor's fence, the romance of spaceflight and the dawning reality seemed counterpoised. To a twelve-year-old boy, "White Sands" had the sound of a holy place, a magic portal to the moon and Mars, worlds still closer to Kepler's Lavania and Burroughs's Barsoom than to future photographs from Apollo and *Viking*. In that time of beginnings, when Heinlein wrote of a boy and his uncle building a backyard rocket and flying to the moon, the vision of spaceflight had the innocence of the early pioneers, who tinkered with rockets in the belief that they would soon be riding them.

Like its cold war counterpart, Kapustin Yar, White Sands suited the adolescent dream. It was a harsh, lonely desert where an isolated few secretly sought the power to break the bonds of Earth. The drive to the edge, having reached the ice cliffs at the literal ends of the Earth, now reached out to "conquer" space. But like the rocket team, awaiting the morning launch, or the vigils of Lowell searching the sands of Mars, the quest for the center, from Kepler to Korolev, had come through a long desert night. With the view from space restoring Earth to the spiritual center, we awaken from what Blake called the "single vision of Newton's sleep."

Coming of Age in the Cosmos

Wriggling from his orbiting capsule for man's first walk in space, cosmonaut Alexei Leonov was born into the starry deep, floating at the end of his umbilical. All about him was a blackness so intense it seemed he could reach out and touch its texture, a darkness so deep that the hair rose on his neck and the flesh crawled on his back. Before him lay the vast curvilinear presence of planet Earth—soft, glowing, haloed against the black abyss. Leonov tumbled slowly in the windless, odorless void, hearing only the rhythms of his body, severed from the mothering Earth—who seemed at long last to have borne a space creature like herself.

Those who have shared Leonov's view of Earth report a pristine clarity, a vividness uncaptured by photographs. Images on film lack the subtle shades, the brightness, and the depth of the living sphere, which bulges out of the blackness. From the blue marbled Mediterranean, all of Europe and Africa sprawl away in soft pastels, innocent of political boundaries. There is an aura of utter reality, of waking from the long, murky dream of man's moment on the surface. One longs to summon back all the Caesars, Pizarros, and Napoleons and put them out there that spring morning with Alexei Leonov, with the radiant arc of Earth floating up like a great leviathan surfacing on the stellar ocean.

The image of Leonov drifting alone in space is the consummation of modern history. For the five-century rise of spacefaring man saw the ascent of the free-floating individual, severed from sources of meaning, terrestrial astronauts adrift in urban bubbles and left to invent their own lives. And humanity itself comes to the end of the age listening like Leonov to the sound of its own breathing.

Though some see our severance from Earth as symbolic of this deepening isolation—the last great Faustian act—others see a communal awakening with the image of a living Earth, the unifying circle of world mythology, paradise regained by the same inflation through which it was lost. In truth, the leap into space stands with the voyage of Columbus as the last act of an old order and the first of a new. Just as the attempt to crown the glory of Europe with the wealth of the Orient relinquished the future to the opposite hemisphere, the image of our fragile, lonely world rising over the dead moon encouraged a rebalance from outer toward inner space, the self-awareness that comes when we realize that the parent is finite and mortal. From such cultural eversions come the new mythologies, the new *Zeitgeister*. One awakens from a dream carrying fresh symbols from the unconscious, the prophet wanders into the desert to reappear with the word of God, and the astronaut reaches the moon to return with a new image of Earth.

Riding his space pod out of the blue planet on a pillar of flame, Leonov was borne from the flat world of our fathers in a fiery rite of passage, marking the end of man's adolescence. If the five-century age of exploration is the collective parallel to the growth of individual ego-consciousness, the astronaut's severance from the source is the final inflated act. From his literal alienation, his pilgrimage into the desert, comes communion with a larger reality. Seeing Earth not as an extension of man, but man as an extension of Earth, we come of age in the cosmos.

If Leonov's odyssey is symbolic of the present transformation, the rise of spaceflight itself is the metaphor of modern history. Just as Kepler's medieval quest launched the modern age, the dreamers of spaceflight sought the rebirth of wonder—the remystification, the reenchantment of the world. Yet in this cycle of inner and outer realities, the promise of spaceflight holds more than the redemption of Mother Earth. Perhaps evolution is a narrowing spiral, seeking a paradoxic union of poles that is itself the Center, each turn bringing a new depth of vision. The longings that bore Leonov into the starry deep sought a new equation of inner and outer space, merging the feminine mysteries of earth and soul with the masculine quest for heaven and spirit.[17] Standing between two unfathomable infinities, we grasp the one only as it reflects the other.

Swan Song

We are creatures of the image. Though Leonov was the first to walk in space, it was a photograph—that of the first American space walk some three months later—that caught the attention of the world. Of the four famous images to emerge from manned spaceflight, it was neither the whole Earth, nor Earthrise from the moon, nor Aldrin poised on the lunar surface, but Ed White, floating over the milky blue vastness of the Pacific with that flash of sunlight on his visor, that best captured the exhilaration, the euphoric liberation of a species come of age. Drifting lazily on his back, somersaulting, pirouetting, perched on the gleaming titanium of Gemini's hull, White had to be coaxed back to the capsule. "I don't want to come back, but I'm coming," he said. "It's the saddest moment of my life."

Tall, lanky, clean-cut, devoted to family and country, White was the all-American astronaut. But this extraordinary athlete who had barely missed the Olympics, the solemn man with a huge smile and gentle, patient eyes, was martyred with Gus Grissom and Roger Chafee when a fire gutted the Apollo 1 module during a rehearsal on the launch pad. White struggled for a sixteen-second eternity to open the hatch, only to be found mired in a web of melted debris, his space suit fused to the floor.

For three astronauts to die suddenly on solid earth at the very hub of NASA know-how, in the midst of monotonous routine, seemed as far from the romance of spaceflight as the austere military funeral—the flag-draped coffin, the widow in dark glasses, the small boy in his best coat and cap, freeze frames in a forgotten issue of *Life*, from that chill January day when Ed White was buried beneath a lacework of bare trees on a West Point hill. His monument is neither the headstone nor the lunar crater that bears his name but the swanlike figure in that famous photo, afloat on the halo of the world.

Laden like ugly ducklings with centuries of self-doubt, we looked back on the water-blue Earth with Ed White and knew that we too were swans.

Whither is fled the visionary gleam?
Where is it now, the glory and the dream?

—Wordsworth,
"Ode on Immortality"

Chapter Two

The Romance of Spaceflight: Nostalgia for a Bygone Future

SOON THERE WILL BE NO ONE who remembers when spaceflight was still a dream, the reverie of reclusive boys and the vision of a handful of men. Most of those who met in ardent little groups in small cafés between the world wars, planning voyages to the moon and planets that they never hoped to witness, are no longer living. And the last lonely youth to lie in a cricket-pulsing, honeysuckle night and gaze at a virgin moon is now in the latter half of his life. On the yellowed pages of boyhood books the silver ships still poise needle-nosed on the craggy wastes of other worlds—on moonscapes bathed in the stark light of some monster planet whose ring-shadowed hemisphere fills the horizon, looming behind space-suited specks who wander across the incandescent night.

The dream had burned beneath the cold and solitary vigils of mountaintop astronomers like Percival Lowell, and in the visions of lone inventors like Robert Goddard. The fantasy had fueled the science fiction of Verne and Wells, of Serviss and Smith and their pulp successors; it filled the monthly pages of Campbell's *Astounding*, the early novels of Heinlein, and the popular science of Ley and Clarke; it radiated from the covers of *Fantasy and Science Fiction*, the paintings of Chesley Bonestell, and the films of George Pal. It was a dream of visible planets impossibly distant, of fantastic alien surfaces, awaiting for eons the beaching of man's boats. It was a vision of steaming Venusian jungles and fine soft days on the green hills of Mars, cooled by coastal breezes from the Great Canal, looking over a far desert where ruins stood half in sand.

It is not that the dream has disappeared; we may in fact be approaching a scientific watershed even more profound than that of Galileo and Newton. But with the coming of spaceflight, as with all change, there was something gained and something lost. Perhaps the public apathy surrounding the space program has reflected in some measure the discrepancy between dream and reality. For though more meaning may lie in one message from the Mars lander than in the most exalted fantasy, the images of spaceflight that proliferated at midcentury arose from oceanic interiors more remote and mysterious than Mars itself. The romance, in short, had a reality of its own. Acquiring its familiar outlines in the pulp subculture of the twenties and thirties, it exploded into mass culture in the late forties and early fifties.

Perhaps I will be one of the last to have known this credulous dream in pre-*Sputnik* form. I was seven when the first American V-2 rocket roared off of White Sands Proving Ground in 1946; I was ten when it boosted a small sounding rocket across the threshold of space, and nineteen when the first artificial satellite shocked the world. The romantic dream of space reached its apogee in those postwar years, when the fantasy of spaceflight and the promise of reality seemed almost in balance. Into this midcentury moment stepped a few writers, artists, and filmmakers who would epitomize the dream of other worlds. Giving final impetus to a science-fiction boom that had been trying to happen since the twenties, they educated the person in the street to the possibility of spaceflight, bringing to mass consciousness the classic dream of the modern age.

Starry Days On the Coast of Saturn

Leafing through *Life* magazine in the last week of May 1944, one found familiar wartime fare: soldiers napping in foxholes amid the rubble of Italy, American boys with their English girls in London's Hyde Park, awaiting the invasion of France. But the tide of the war had turned, and in the time between Hitler's retreat and the coming of the cold war, a fresh breeze blew across America. One sensed it in the record success of *Oklahoma!* with its aura of youth, hope, and new beginnings, and in the spate of plays and novels set in the sunnier days of turn-of-the-century America. But like the paradox of adolescence, torn between the security of home and the promise of the world, the new optimism often sought the simpler past within a wondrous future. Thus that May 29 issue of *Life* included not only ads out of Currier and Ives and a long piece on *Oklahoma!*'s lyricist, but a large, singular painting, leaping out in vivid color amid black and white pages, depicting Saturn as seen from the surface of Titan, its largest moon.

Since Titan is the only satellite in the solar system with an atmosphere, the giant Saturn looms low in a dark blue sky like an alien ship, a thin, gleaming crescent bisected by the glowing edge of its rings, afloat between jagged cliffs that jut from a frozen sea. Warmed by the distant sun, the rocky cliffs and scarps rise sheer into the cobalt sky, casting a dark shadow on the icy sea. There is an eerie beauty in the incongruity of light. One feels that a storm has passed on a late November afternoon, yet the sky is specked with stars. A hint of dawn lights the far horizon; and beyond a lofty pinnacle, out under the glow of the great crescent, lies a distant patch of noon-day plain. The painting could pass for a photograph in the era of *Viking* and *Voyager*, but on the eve of the invasion it was one man's vision of the future, later recalled by astronauts, rocket men, and science-fiction fans as their first encounter with a dream about to take wing.

On the same page are two smaller views of Saturn as seen from its farther moons Phoebe and Iapetus. Since neither moon has an atmosphere, the distant Saturn floats in a black sky on the phosphorous mist of the Milky Way. The absence of haze and dust gives the landscapes such clarity, the light such purity, that rocky features remain sharp in the distance. The scene has the feel of a great indoor arena, of a room so large that one cannot see where the dark ceiling begins. The paintings were intended to show the changing aspect of Saturn as seen by a traveler hopping toward it from moon to moon. On the next page one finds a gargantuan segment of the planet as seen from its near moon Mimas. With its stark slash of ring shadow, the yellow-banded Saturn balloons into the blackness. Sheer cliffs and jagged mesas meander the red-brown desolation of Mimas, strewn with rocks and ragged craters, stretching away to the hazeless horizon that slices across the monster Saturn.

The six paintings were the work of an architect who spent nearly three decades helping to design landmark buildings across the United States. He went to Hollywood in 1938 at the age of fifty and became the highest paid special-effects artist in the business. The notion of rendering astronomically accurate views of Saturn came not only from his architectural focus on realistic detail but also from his skill in achieving the continuity required for matte paintings representing different camera angles on the same scene. In 1944 he took the paintings, unsolicited, to *Life*, which bought them immediately. Though crude images of spaceflight had been the staple of pulp covers in the thirties, the leap to a larger audience was almost completely the work of this one man. Among those who grew up in the forties and fifties and later became prominent in space fact or fiction, there are few who would not cite the art of Chesley Bonestell as a significant personal influence.

His photographic realism gave the readers of *Life* a new perspective on the night sky. According to the best science of the time, this would have been their view had they actually stood on the moons of Saturn. Bonestell did for the heavens what the microscope did for our perception of life, opening worlds within worlds, inviting adventure, converting points of light into real places. "You may roam about here," his paintings seem to say. "This mysterious island, fresh from creation, is made a place by your mere presence." Bonestell put tiny space-suited figures in most of his scenes, for which one could search like a signature, a perspective reminiscent of the sublime landscapes of nineteenth-century romantics. In painting the planets of other suns, where the setting of a red giant might span the whole horizon with an ethereal arch of deep orange fire, Bonestell expressed his faith that light-years would not forever imprison us in the solar system; for in the late forties, even Mars seemed so remote that whoever could touch it would surely reach the stars.

For those who grew up with Bonestell's painstaking accuracy in light, shadow, perspective, and scale, the reality of spaceflight seemed a foregone conclusion. In March 1946, *Life* published twelve more paintings by Bonestell depicting a hypothetical flight to the moon; and in the next two years, his space illustrations appeared in *Scientific American*, *Coronet*, *Pic*, and *Mechanix Illustrated*. For the October 1947 issue of *Astounding Science Fiction* he did the first of his many covers in that genre. But it was *The Conquest of Space*, a book published in 1949 with text by German rocket expert Willy Ley, that was primarily responsible for bringing Bonestell's planetary landscapes to a new generation of spaceflight enthusiasts.[1] It included the paintings from *Life* and many more: fairy-tale landscapes laced with castlelike rocks carved by drifting dust; lava spilling over the icy cliffs of Jupiter; the rocky green hills of Mars, rolling like the coast of Maine along the great canals; and on the near moons of enormous planets, knife-edged peaks and needles of rock stabbing into a star-filled sky. Together, the scenes suggest a kind of cosmic shoreline, a composite of stark and eerie beaches on the near edge of the starry deep.

It is not surprising that the most popular of Bonestell's paintings, the view of Saturn from Titan, resembles a rocky coast in the frozen reaches of the far north. For the image of the beach is not incidental. Like the dream of space-flight itself, the appeal of *The Conquest of Space*, which went through four printings in the first three months, owed much to the archetypes of the seashore. My discovery of that book on Christmas night of 1950 had an impact very similar, in fact, to my earliest memory of the beach.

The Only Real Place

I was four years old in 1942 when the army sent my father to Fort Ord on California's Monterey

Peninsula. We left a dreary flat in the gray mist of San Francisco for a sunny cottage near the cypress-lined, white-sand beaches of Carmel. As if to ritualize this rebirth, my mother took me for a walk that wound through a dark grove of those great brooding cypress—leaning and reaching with their gnarled, windswept limbs, growing ever more foreboding—until the path opened suddenly onto a long stretch of pure white sand and a vast expanse of silver-blue water that sparkled and shimmered to the edge of the world. It was my first waking encounter with the Pacific Ocean. I ran barefoot over the hot sand, stopping at a safe distance to gape at the bellowing breakers, feeling the cold foam on my feet. As vivid still as the smell of ice plant on the dunes, it is a moment burned into memory, like an astronaut's image of Earthrise from the shores of the moon.

We went often, through the dark trees to the sunlit beach. I built sand castles while my mother sat on the grassy bank with the salt-kelp breeze in her hair, watching boat specks on the horizon. Her death, shortly after we left Carmel the following year, seems to have merged my sense of the mother with that of the ocean—Great Mother of all, mystery of origins, milk of the world; the Good Mother, nurturing a silent undersea fantasy of living things; the Dark Mother, swallower of worlds, the black sea-bottom of death itself, strewn with *Titanics*, digesting Atlantis and Lemuria.

The epilogue came a few years later at a summer camp in the high mountains. I awoke one night in a sleeping bag under a wilderness of distant worlds, recalling Asimov's story about a planet with six suns, where "Nightfall" occurs but once every 2,050 years and the sudden appearance of a soul-searing canopy of stars plunges civilization into chaos. Gazing out into the immense ocean of light I reexperienced my encounter with the Pacific, though there was no odor of ice plant on the breeze, no sound of breakers nor wind in the cypress, only the silence of those trillion worlds, waiting, eyes within eyes, coming through a million lifetimes to meet mine—which glanced away, struck with what we all come to know: that each of the unfathomable immensities—Mother, Ocean, Death, and Stars—share the barrier between known and unknown, enfolding the familiar world like the pre-Columbian gods and monsters, bounding all beginnings, all ends, all meaning.

Perhaps it was gods and monsters, not gold and glory, that inspired young Cristoforo Columbo on the shores of his boyhood Genoa, gazing out on a sea that encircled the known world like the night sky—a fathomless enigma, fading off into forever. Men once looked out over the melancholy wilderness of water as we now look to the stars, knowing it to veil some great mystery of unknown size and origin. Though the sea no longer bounds the universe, it remains a vast, inscrutable presence, growing darker and deeper in the distance, the darkness of a world before man, unchanged through eons of continental evolution, yet ever restless, relentlessly pounding the land, through all lifetimes. The rush of the surf echoes the ancient Earth—the wind in the once great forests, the thunder of free-running herds—while the sea alone remains truly free, the last untamed remnant of Earth's tempestuous youth. And out beyond the breakers abides the silent face of the Great Mother, an effervescence of light, flashing like countless suns.

Though the archetype of the ocean shapes the aura of Bonestell's paintings, the root metaphor is more precisely the shoreline itself. The interface of known and unknown, civilization and wilderness, conscious and unconscious, the beach is that narrow band of equilibrium where the city meets the sea. To go down to the sea-scented shore on a cold, gray day and wander amid the wrack and debris of both worlds, to sit on a half-buried whiskey box, watching the birds dip and hunt with their small sad voices, is to enter sacred space, to walk the razor's edge between time and eternity, matter and spirit, isolation and communion. The seashore is a sanctuary, the eye of the storm, where our polarities are momentarily balanced. It is where the temporal realm of the hot street—even the run-down hot-dog stand—is bathed in transcendent energy, touched by the breath and pulse of the sea.

The transformation is reciprocal. With each mortal breaker the eternal sea dies a momentary death, descending into time as it licks the sands to the soft cries of gulls. Yet the beach is a place of rebirth, where each wave erases the tracks of life and time, leaving the broad sand flats gleaming like glass. It is a place where false selves are shed and companions transcend their separateness. It is a holy place, perhaps the only real place.

Islands at the Edge

The same paradoxic polarities are at the core of Bonestell's appeal. To feel the lure of the seashore is to know why the near planets still beckon, even when the hope of inhabitants is lost. By 1950, humanity itself had become the city on the shore of the cosmic ocean—a growing cancer of disconnected egos living in the shadow of the bomb. For an eleven-year-old awakening to that larger reality, the barrier between known and unknown first encountered on the coast of Carmel now became the pristine beaches of Bonestell's planets (it is fitting that Bonestell himself lived in Carmel). Like the seashore, his alien landscapes domesticated the transcendent while elevating the mundane, familiarizing the mysterious and mystifying the familiar. His most popular painting, the view from Titan, with its giant Saturn afloat in a blue sky over sunlit peaks, seemed the ultimate marriage of strange and familiar.

In the hidden heart of science fiction had always been the hope that the moon and planets promised a personal adventure akin to my encounter with the Pacific—a transitory, enchanted moment when man, like Fitzgerald's Dutchmen, would hold his breath before the fresh green breast of some radiant new world. At that moment, the cosmos would become a place. This was the promise of Bonestell's beaches—that those peaceful points of light in the night sky are *places*, that virgin rocks, asleep for a billion years, await my touch no less than the cup that sits here before me, that there is a "transcendent mundaneness" abiding in parallel time and space, one that somehow merges the cosmic and the personal. A Bonestell moonscape is a sacred place at the edge of the known world—an altar set before the barrier, a piece of the mundane bathed in oceanic mystery.

We think of the boundaries of the known, the outer rim of our reality, as somehow harboring the answers to our "Why" questions. Whether it be Aristotle's geocentric spheres, Columbus's ocean-sea, or our own space-time continuum, we conceive of this Larger Context as ultimately separate from and alien to our everyday experience simply because it is assumed to mask the unknowable, the meaning of all meaning, as inaccessible to us as cosmology to an ant. To encounter the blue skies of Bonestell's Titan, or to find that the reddish, rock-strewn desert of Mars looks hauntingly like the American Southwest, suggests that the near edge of the Larger Context is a reality as familiar as our own backyard. Just as the beach is perceived as the edge of an otherwise boundless sea, Bonestell brought the edge of infinity out of the abstract and into the realm of direct experience.

The two realms, abstract and direct, are quite different. In one, red is a designated wavelength of light; in the other, red is a color. In one, the Earth rotates on its axis and revolves about the sun; in the other, the sun rises and sets. The photographic realism of Bonestell's cosmic beaches brings the realms together at their point of tangency, allowing one to experience the heliocentric reality from a geocentric perspective, giving the Larger Context a tactile immediacy. The paintings had a disorienting effect, similar to that experienced by astronauts seeing Earth from the moon. Standing beneath the blue swirl of Earth, said Gene Cernan, "I had to stop and ask myself, 'Do you really know where you are in space and time and history?'" To believe, with a foot in each world, that one can retain an earthly reality yet stand on *another* ground under *another* sky has the transforming impact of an out-of-body experience, or an encounter with the doppelgänger, a duplicate self.

The lack of inhabitants allows Bonestell's barren astroscape to become an extension of oneself. The image suggests a colossal stage set or a giant playroom, a toy world of one's own. There is a timeless stasis about these bright islands, offering safe passage through the void as one might ride a cozy car through the Tunnel of Horror. A virgin purity, untouched by everyday life, not only makes the place one's own but beckons in the way that tiny islands lure the canoeist. The scenes, like the islands, invite one to become a world unto oneself. It is the narcissistic fantasy of infusing whole societies, planets, or universes with one's own nature and agenda—an utter absence of personal limits. But unlike the way of the mystic, the ego is not sacrificed; it is merely romanticized. For Bonestell's landscapes, like the canoeist's island, remain pieces of the real world, filled with those craggy shadows, hidden caverns, and pristine horizons that are natural to the romantic imagination.

Such romantic flights from personal limits were more than the marks of innocent youth. For as long as spaceflight was far from reality, one could immerse oneself in the subject with the intensity and resolve of real exploration—the exhilarating sense that one stood very near the leading edge simply by combing musty libraries or gluing balsa models in a cluttered garage. Such mundane pursuits seemed to merge with the ultimate quest and adventure on its realistic timeline. Lying in the hammock and gazing at the moon, viewing the larger craters through Dad's binoculars—so close, yet so impossibly remote—one reached another world by the same means to which even the most heroic adventurer was then bound.

Yet one took the risks without leaving home, gained the world without losing the soul, pursuing the real adventure in a context as warm and secure as reading comics in bed while mother nursed the common cold. For space travel, like all mythic visions, was paradoxic, containing at its core the very polarity peculiar to most adolescent males. The parental womb—the inner solipsistic world of childhood—became the secret spaceship, while the external world was removed to outer space, where one's omnipotence was safely assumed. One eluded the looming world of adult responsibilities, the messy arenas of peer conflict and opposite-sexed enigmas, withdrawing to a larger realm that encircled those lesser things. There one finds Bonestell's planets inhabited by other childlike beings, or by benevolent and omniscient Cosmic Parents. Substitute for the parental womb the technological society, where the illusion of infinite leverage compensates for a state of abject dependence, and one has an inchoate hypothesis with which to explain the rise of science fiction itself over the past century.

Before I sentence all science-fiction fans to schizoid adolescence, however, I should note not only that all neuroses merely exaggerate facets of normal behavior but also that the progressive/regressive polarity is common to all transition, individual and collective. The adolescent is in fact a microcosm of the modern condition, wandering the beach between two worlds. His isolation is that of humanity itself, a species cut off from its traditions, its instincts, its animal forebears, even its home planet, while riding the momentum of history like a runaway train.

Ironically, while Bonestell's island beaches are stepping stones to the discovery of higher life, they also intensify this sense of isolation. Looking out over the ocean or gazing back on the home planet, one feels the solitude of humanity, a species cast up from the ancestral seas and forests, standing alone at the leading edge, like an old man who has lost all those from whom he issued. Just as whole cultures reach a zenith of awareness and creativity when old and new are in polar tension, so awareness itself may be the product of the "bicameral mind," the hemispheric tension which allows the stereoscopic, three-dimensional thinking that is consciousness. This reflexive self-awareness, the ability to think about thinking, differentiates us from other animals. It is the source of all good and all evil, all cruelty and all compassion. It is this reflexive tension that pervades the shoreline, terrestrial or cosmic. It is the fragile polar balance that defines the human condition, bringing both heightened awareness and the concomitant solitude of separation. On the beach, faces flicker in the firelight—creatures marooned on the shores of evolution, gathered to mourn and to celebrate their loss of innocence.

Destination Moon

In one of Bonestell's paintings for *The Conquest of Space*, a sleek silver rocket towers on its gantry against the stars. Men move about in light and shadow as though huddled around a great silver flame, a signal fire warming the night at the edge of the cosmic ocean. In a second painting, the same ship rests on its fins in a valley of the moon against black sky, sunlit mountains, and the distant Earth. Light radiates from the hatchway as the crew explores the site, a rugged outpost on the frontier of evolution. In the hearthlike glow of these scenes lies a communal longing, life reaching out for some indiscernible secret, some wormhole to the long forgotten Source, the heartsong of creation.

The paintings, along with additional moonscapes, became the basis for visual effects in *Destination Moon* (1950), the film often credited with launching the cinematic science-fiction boom of the fifties. Film historians traditionally note that *Destination Moon* was the first science-fiction film worthy of the term since *Things To Come* (Britain, 1936) and that its success led to the boom that followed. Written by Robert Heinlein and Rip Van Ronkel, directed by Irving Pichel, and produced by George Pal, with panoramic matte paintings by Bonestell, *Destination Moon* depicts a trip to the moon, true in almost every detail to the scientific projections of the time. The result is a docudrama surprisingly close in particulars to the flight of Apollo 11 nearly two decades later. Although it was Eagle-Lion's top moneymaker for the year, earning enthusiastic reviews and an Oscar for special effects, the film has not worn well. Film historians lament the absence of plot, women, and depth of character, dismissing the film as dated and dull.

The problem with this kind of film history, of course, is that it lacks historical perspective. Not only do many film critics seem bound to secular realism, but most of the writers are too young to have seen *Destination Moon* when it was released. When the long silver ship set down on the surface of the moon and the crew descended in silence, it was as though irrelevancies dimmed and essentials came clear. Viewing the first realistic depiction of a visit to another world (also the first in color) had an effect similar to the televised landing of Apollo, or to the dazzling NASA footage projected onto the five-story Imax screen in *Blue Planet* (1990). The one previous attempt at an accurate representation of space travel, *Woman in the Moon* (Germany, 1929), was far less compelling, not only because its moon sequences reverted to fantasy but also because it was silent at a time when most films had incorporated sound. The degree to which *Destination Moon* escaped hack concessions to Hollywood formulas is a credit to Heinlein and Pichel. With the help of Bonestell and Pal they conceived a prophetic film, an ode to the romance of spaceflight that retained, even in its pedestrian moments, an aura of transcendence.

But was it responsible for Hollywood's spate of science-fiction films? In truth, the producers of almost all of the significant SF films released in its immediate wake had conceived their ideas and purchased their properties prior to the contracting of *Destination Moon*. Along with the surge of SF magazines, the notion of filming serious science fiction was in the air in the late forties. Although the optimistic and visionary *Destination Moon* was virtually first in the cycle, its impact was confined to demonstrating that expensive SF could be profitable.[2] Most succeeding SF films fell into the Gothic mode, embodying the mutated monster and evil alien themes of *King Kong* (1933; rereleased in 1952) and *The Thing* (1951).

In the end, *Destination Moon* was less a beginning than a culmination. Epitomizing that balance of fantasy and reality that characterized the dream of spaceflight itself at that transitional moment, it was the confluence of three historic careers: Bonestell, the first realistic space artist; Heinlein, whose gritty realism depicting life in space had made him the first science-fiction writer to break into a mainstream magazine (with his *Saturday Evening Post* stories in 1947); and Pal, the first science-fiction filmmaker to show concern for scientific credibility. Pal's bias toward realism had been evident in his Puppetoons, which were not only the first socially minded cartoons but also first to use stop-motion photography of real figures in place of animation.

Pioneers in fantasy and innovators in realism, the three men surfaced at a time when the developing facts of spaceflight had begun to reshape the nature of the dream. *Destination Moon* perfectly bridged the romance of the dying pulps and the coming realities of Apollo. After *Destination Moon*, in fact, spaceflight on film suffered a fate similar to that of the real thing after Apollo. In the public mind, the deed had been done. Pal retreated to traditional science fiction, turning first to the destruction of the Earth, in which the same sleek ship from *Destination Moon* became a planetary Noah's Ark (*When Worlds Collide*, 1951, with artwork by Bonestell), then to alien invasion (*War of the Worlds*, 1953), time travel (*The Time Machine*, 1960), and finally back to fantasy from whence he came. Pal tried a second docudrama (*The Conquest of Space*, 1955), but its success fell short of *Destination Moon* to about the same degree that public interest in the space shuttle fell short of attention to the moon landing. Heinlein attempted another script (*Project Moonbase*, 1953), expanded from an unsold TV pilot and rewritten by the producer, with results that forever soured Heinlein on Hollywood. His later books, like Pal's films, moved away from the pure extrapolative realism of earlier SF. Bonestell, on the other hand, moved toward the emerging realities with his illustrations for Wernher von Braun's famous *Collier's* articles in the early fifties. Thus the romantic vision receded with *Destination Moon*, where it abides like an old daguerreotype, the finest and final hour of a dream now faded and transformed.

Yellow Brick Road to Redrock Corners

Perhaps the images of unencumbered human flight that preceded the realities of the airplane included winged torsos hovering high above trees and houses, soaring like eagles over hill and forest, swooping like hawks down rivers and valleys. During the first half of the twentieth century, the romantic vision of coming adventures in the solar system compared in spirit to just such reveries. The depersonalized, mechanical reality of astronautics was never a part of the dream. When we finally went to the moon it was not on a wing and a prayer but on a pyramid of mathematics and technological expertise. Every move was part of an exhaustively detailed script. Should the left or the right hand pick up this piece of equipment on the moon? Should the knuckles point up or down? With simulations more novel and rigorous than the actual flights, the moon's only surprise was that it held no surprises. In our business, said Apollo 11 astronaut Mike Collins, "boring is good because it means that you haven't been surprised, that your planning has been precise and your expectations matched."[3] "Adventures," said polar explorer Amundsen, "happen to the incompetent."

What faded in the years following *Sputnik* was not the dream itself but its naïve forms. The astronomical interests of most rocket pioneers were superficial, romantic, and unscientific, while few professional astronomers took spaceflight seriously. Even rocket research had once been romantic. Robert Goddard's telemetry in Roswell consisted of a pair of binoculars, an old alarm clock to drive a recording drum, and Esther Goddard's movie camera. She was photographer, secretary, and parachute seamstress, among other titles. It was an intensely personal quest. Goddard and fellow rocket pioneers Konstantin Tsiolkovsky and Hermann Oberth all devoured the science fiction of Verne and Wells, with Oberth's interests extending even to the occult. Like the masses who panicked during Orson Welles's 1938 "War of the Worlds" broadcast, early spaceflight enthusiasts not only took seriously Percival Lowell's insistence on Martian canals but envisioned a lush, tropical Venus and solid, surrealistic surfaces on the outer giants. Drawing on the primitive state of astronomy, pulp fantasies seemed to confirm that if one could somehow hurl oneself off the Earth, one would encounter myriad yellow brick roads to an infinity of Emerald Cities.

Today the moon and planets are not only inhospitable but have undergone a certain desacralization. The romantic, mysterious, inaccessible moon that made the water silver, the swollen tangerine Allegheny moon, the "ghostly galleon tossed upon cloudy seas," the moon that only cows in nursery rhymes could jump over—that moon is gone. The once holy ground of myth and magic is now a barren, hostile desert.

It has been suggested that there are seven zones of human experience: (1) the area of sensation immediately touching the skin, (2) the area within two or three meters in which most social interaction takes place, (3) the maximum area of social interaction, reaching out a few hundred feet, (4) the area that extends as far as one can see or otherwise gather information from any one location, and (5) an irregular and varying area made up of all the zone-four areas that a person experiences during a lifetime. Beyond these five natural zones are two conceptual constructs: (6) the surface or biosphere of the Earth and (7) the universe as far out as one can conceive.[4] Our extraverted, Newtonian culture has viewed the outermost zone as the realm of transcendence, the literal locus of ultimate answers to all our "Why" questions. In the course of one lifetime, this zone has expanded at least three-hundred-thousandfold. With the discovery in 1921 that our galaxy was not the sum total of existence, the solar system moved from zone seven to zone six. In the new cosmos, the planets seem to lie less on the near edge of infinity than on the far edge of the Earth.

Yet just as the beach is bathed in the aura of the sea, Bonestell's planets could still symbolize that unfathomable immensity. It was not until we viewed the Earth from space, left our spoor on the moon and Mars, and intruded on the time-less solitude of the outer planets that we began to experience what we had known. The rolling gray lunar hills have belied the jagged, craggy wonderland of *Destination Moon*; and the rocky red desert of Mars is more like a spherical New Mexico than the home of Wells's doomsday machines. Now a part of "where we are," of zones four and five, given the disorienting impact of electronic media, they become Grayrock Junction, Redrock Corners, and Gasball City, associated less with the Land of Oz than with Steinbeck's flat-country truck stops—those dilapidated diners with a gas pump in front, where flies strike the screen door with little bumps and drone away.

To conceive of the transcendent requires a symbol. One cannot worship "God"—a word, a vague feeling, an intellectual abstraction; one needs an image: Jesus, Buddha, a bearded man in the sky, a painting, a statue. Yet the natural tendency is for the image to literally become God, and the larger, elusive feeling that empowered the symbol fades, eclipsed by a host of this-worldly connections. Finally demystified, the statue reverts to its status as mere artifact. Thus every symbol contains the seeds of its own desacralization. The millennial nature of Christian theology generated the idea of spiritual progress, which spawned the notion of salvation through success in this world, which led to the secular idea of material progress, which in turn began to desacralize Christian theology. The last stage of this process is fundamentalist dogma, in which symbols have lost their numinosity and have degenerated to mere signs.

The Cartesian-Newtonian worldview, which has deferred its "Why" questions to the empirical edges of space and time, faces a similar cycle. Like the instruments of Christianity, discoveries at the leading edge of science desacralize the very things that compelled the quest in the first place. The sense of wonder surrounding those objects suffers the fundamentalist fate: the mythic moon becomes a wilderness of cinders and ash, and the red star of evening becomes a barren, rocky desert. The bright light of science dissolves the mysteries that animate its objects; the observer alters the observed, and the aura of wonder recedes with the horizon. We pine for the lost images demystified by modernity, and the innocent dream of spaceflight joins the romance of the railroad in coffee-table nostalgia.

The Archimedean Point

Now that we have touched a heavenly body, the rockets themselves no longer seize the imagination. In a field near the Houston Space Center, the last Apollo lies like a beached whale amid a trickle of visitors. It was once enough merely to escape the Earth, but now the core motive comes clear: cosmic communion. "Even the traveler's mind," wrote a post-Apollo poet, "now shoots quicker than a gecko's tongue beyond the sun for the sweet stars, thrilled by demons, by impossible virtue and impossibly wise old men."[5] Like Bonestell's planets, the moon once as remote as the stars; just to stand on it would unmask the night sky, rendering the whole cosmos as accessible as the worlds of science fiction. But the realities of reaching the moon drove science fiction from the "hard," mechanical extrapolations of *Destination Moon* to the "soft" phantasms of *2001*, *E.T.*, *Starman*, *Cocoon*, and *Contact*. Forced inward to the promise of Christlike aliens, paranormal realities, and mystical resurrections in space, science fiction no longer pretends to paint the near future.

Perhaps the new dream is less naïve than the old. Arthur Clarke has noted that we tend to overestimate what we can do in the near future and grossly underestimate what can be done in the distant future. This is because the imagination extrapolates in a straight line, while real events develop exponentially, like compound interest. Perhaps communication is not limited by the speed of light; perhaps there are "wormholes" in space-time; perhaps we will receive some mind-altering message from superior beings. Such hopes were the subject of Clarke's 1968 film *2001: A Space Odyssey*, the first film of any consequence about spaceflight since *Destination Moon*, and the only other such film scripted by a major science-fiction writer.

2001, in fact, capsulized the transition from the old dream to the new. When the ape-man spins his bone tool into the air, where it dissolves to a wheeling space station complete with pay phones and a Howard Johnson's restaurant, the message is that the colonization of space will be no more than an extension of man's tool-making nature. In the first half of the film space is presented as essentially more of the same, a bland, anonymous world set to the "Blue Danube" waltz, a Howard Johnson's in the sky, where means remain ends in themselves. The second half of the film moves from the "how" of spaceflight—the rockets, space stations, synthetic foods, and supersophisticated computers—to the "why," symbolized by the inscrutable aliens who transform the astronaut into the mysterious star-child, drifting toward Earth to be born. The two halves of the film depict the shift from the old dream, terminating in the simple act of escaping the planet, to the new dream of discovering some clue to our meaning and destiny, of finding life, and of launching the long journey in which man may evolve into a new galactic species.

What, then, is the lure of the near planets? Why dedicate a life to a landing on Mars? Like the spice islands envisioned across Columbus's ocean-sea, Mars was once the mystery of the cosmos incarnate. Now it is only the near edge of the night sky. Yet the red sands of Mars, now part of our reality, still lure us to the shore of the cosmic ocean.

Although Mars is no longer shrouded in mystery, it remains what Hannah Arendt called an "Archimedean point" (it was Archimedes who said that with a long enough lever and the moon as a fulcrum he could move the Earth). Applying the term to the tendency of modern science to substitute its heuristic constructs for direct experience, Arendt suggested that man increasingly encounters only himself. Viewing spaceflight as an attempt to reach a literal Archimedean point, one that must always require a still further point, she saw the leap into space as a flight from the human condition. But as the vision animating all forms of exploration and discovery, the Archimedean point *is* the human condition.

Physicist Philip Morrison tells a story about the Bushmen of the Kalahari, Africa's last society of hunter-gatherers, who forever move about the desert, carrying what little they possess, living in bands of extended families, each staying within a region about the size of Los Angeles County. They meander through life, "stopping now here, now there, to sleep in a kind of nest, to try the fruit of this tree, to scratch up that waterhole," or to meet for a "ritual encounter with their wandering friends." Their wants are so well controlled and their skills so well developed that they need not work any harder. Their one need,

> as they wander through the cool mornings, the cool evenings, and as they rest in the heat of the day, is to know exactly where they are. They discuss it always. They note every tree, they describe every rock. They recognize every feature of the ground. They ask how it has changed, or how far it has been constant. What story do you know about this place? They recall what grandfather once said about it. They conjecture, and they elaborate; their minds are filled; their speech elaborates exactly where they are. You see they have built an intensely detailed, brilliant, forever reinvigorated internal model of the shifting natural world in which they find their being.

Morrison suggests that our language, myth, ritual, tools, science, and art are all symbolic expressions of a "grand internal model" that every human makes, and that is always in need of completion.

The essence of human exploration is the attempt to fill in the margins of that model so that it will not "fade off into the nothing or the nowhere."[6]

All such terms—Arendt's Archimedean point, Morrison's grand internal model, space writer Frank White's "overview effect"—are aspects of the reflexive thinking that is the hallmark of our species. If the essence of exploration is to touch the boundary—the beach, the mountaintop, or the moon—the core of the human condition is the attempt to see the self in context. To stand on the moons of Saturn and see the Earth in perspective is to act out the unique identity of our species.

The Star Thrower

That it is humanity's fate to live alone at the leading edge was a point often made by evolutionary biologist Loren Eiseley. In "The Star Thrower" he describes a wave-beaten coast, littered with the debris of life—upended timbers, "sea wrack wrenched from the far-out kelp forests," and long-limbed starfish strewn everywhere, "as though the night sky had showered down." A hermit crab is tossed naked ashore, "where the waiting gulls cut him to pieces."

Along the strip of wet sand "death walks hugely and in many forms. Even the torn fragments of green sponge yield bits of scrambling life striving to return to the great mother that has nourished and protected them."

> In the end the sea rejects its offspring. They cannot fight their way home through the surf which casts them repeatedly back upon the shore. The tiny breathing pores of starfish are stuffed with sand. The rising sun shrivels the mucilaginous bodies of the unprotected. The seabeach and its endless war are soundless. Nothing screams but the gulls.[7]

Among the competing collectors, clutching their bags of beautiful voiceless creatures, was a human figure framed beneath a distant rainbow, spinning starfish far out over the surf. The star thrower, "whose eyes seemed to take on the far depths of the sea," represented for Eiseley the furthermost reach of humanity. Shipwrecked on the shores of evolution, he is unique in his compassion. He is the catcher in the rye, protecting the innocent from the abyss, doomed eternally to explore the margins, chasing his reflection down an infinite regress.

Like the immigrant wife, alone on the howling Kansas prairie with her Sears catalog and her secret dreams, those who settle the red sands of Mars will know that some of their roots must die on that barren shore. Gazing back on the soft blue dot in the Martian sky, perhaps they will dream of dance music drifting over a moonlit lake, of twilight talk in a turn-of-the-century town, or the dark wet soil of the sunflower forest from whence we emerged like life from the sea. But pioneers on the plains of Mars will no longer debate whether that pale blue point is the center or the edge, for the center, as Joseph Campbell has said, will be man himself, finally aware that he lives in the stars.

The imperative to see the self from afar, to see the present from some external point in the future, is neither Promethean, as Newtonians assume, nor Narcissistic, as Arendt suggests, but closer to the task of Sisyphus, who was condemned to roll the rock up the hill only to have it roll down and eternally begin again. For the succession of Archimedean points—whether metaphoric, as in art, or literal, as in science and exploration—spirals through history. Though we return always to the same point, there is a new perspective with each cycle, and we seem to know the place for the first time. The process is unending, but we *are* process; and our soaring aspirations are finally cathedrals of the mind. For latter-day Argonauts to return with the poster of the whole universe would in fact be a form of idolatry. Were we to awaken from the dream of space-time we might long for some eternal star thrower to return us from the center to the edge. For it is not the treasure but the voyage itself that is the central project of our species.

Again and again we come through the dark trees to the Pacific: the lookout on *Pinta*, spying a hint of white sand cliff in the moonlight; Amundsen, on a bleak, windswept plain of ice, reaching the pole with his few surviving dogs; and Apollo 8, in the hush of Christmas Eve, floating over the mountains of the moon. This was the promise of Bonestell's vision, that people from Earth would one day flow into the ancient river valleys of Mars, down the gorges four miles deep, out over desolate, wind-torn plains, out to the ice seas of Europa, the yellow skies of Titan, and the Great Wall of Miranda, out into the ocean of light, to those worlds within worlds where the star-children wait.

We shall not cease from exploration
And the end of all our exploring
Will be to arrive where we started
And know the place for the first time.

—T. S. Eliot,
"Little Gidding"

Chapter Three

Seeking the Center
at the Edge

POISED ON THE LAUNCH PAD and towering thirty-six stories against the stars, the Apollo-Saturn rocket seemed unearthly in the wash of floodlight, glowing icy silver-white, like the moon above it. A half-million pilgrims had made their way to the mosquitoed marshlands of Florida's Merritt Island, spending the night on the beach in cars, tents, and trailers, awaiting the early-morning launch that would put man on the moon. Along the grassy dunes and desolate moors, onlookers stood in the soft whine of the night wind, the children of four hundred million years of land creatures on this planet. Life, which crawled out of the sea those eons ago, would now climb out of the white-capped ocean of air, cling to a barren rock, and fall back to earth. For one brief moment, we would be creatures of the cosmic ocean.

The lure of the night sky is older than the voyages of Jules Verne, older than Greek tales of winged flights to the moon, older even than the pyramids, which were aligned with the pole star so that Pharaoh might reach it in his sky-traveling boat. The call of the cosmic ocean haunts the high mountains and remote seashores, where the misty river of the Milky Way arcs across a fathomless dome of sand-grain stars—the vacant stare of creation, lying ever behind the painted face of day. "Some part of our being," said Carl Sagan, "knows this is from where we came. We long to return."[1]

Like those pilgrims camped along the beach, listening in the night to the pulse of creation, we live on the shore of the cosmic ocean, riding our wisp of blue and white like mites on a floating leaf, in the whorls and eddies of a great galactic reef. Adrift like an ill-fated liner with her lights ablaze in the North Atlantic night, the lilt of her music faint on the icy wind, we are the ballroom innocents of Spaceship Earth—frail seed of life itself, afloat for an instant on the surface of forever.

A Wondrously Beautiful Thing

Apollo-Saturn is silhouetted on the horizon against a dawn sky, visible across dreary flatlands that stretch eastward to deserted launchpads along the sea. The landscape recalls settings from old science-fiction films—the lonely stretch of desert road under a bleak sky, the gray impermanence of brooding seacoasts, stark lunar surfaces—metaphors for the homelessness of modern man. Apollo 11 seems in fact the consummation of modern history—the ultimate hubris, the last great phallic act of Faustian culture. On its altar of steel and concrete, the Saturn V rocket is the icon of infinite force and mastery, man's collective erection, built to seed the cosmos. It is the hotrod of an adolescent species bent on tearing itself from Mother Earth, bolting the Garden without God's consent.

On Pad 39A, vapors stream furiously off the side of the ship, loaded with enough fuel to fill ninety-six railroad tank cars. Sixty feet taller than the Statue of Liberty, an aircraft carrier set on end, the three-thousand-eight-hundred-ton Apollo-Saturn, with its fifteen million individual parts, is more finely tooled than an exquisite watch.

Perched high atop this cylindric skyscraper, the astronauts lie patiently through the countdown. Were the bomb beneath them to explode, the ball of flame would consume two-thirds of a mile, with a rain of debris eight miles wide.

"T minus fifteen seconds. Guidance is internal. Twelve, eleven, ten, nine, ignition sequence starts, six, five"—the voice of public affairs officer Jack King comes through the loudspeakers at the Cape and half a billion television sets—"two, one, zero, all engines running. Liftoff! We have a liftoff, thirty-two minutes past the hour." Orange and white billows of flame gush out hundreds of feet over the ground, engulfing the base of the ship. Four twenty-ton restraining arms release room-size grips, exploding away with a rain of ice that has formed on the liquid fuel sections. The automatic camera on the gantry tower records a great white wall beginning to rise— inches, feet, two vertical black stripes forming a U, then S, A, passing rapidly now, followed by a wall of fire, lifting six and a half million pounds, gulping fifteen tons of fuel a second, the eight-hundred-foot tail of flame whipping white-hot as the sun. Consuming as much oxygen at the moment of liftoff as a half billion people taking a breath, cooled by water cascading at fifty thousand gallons a minute, the Saturn V rocket rises with the force of a hundred thousand locomotives, burning five million pounds of fuel in the first 150 seconds, getting a full five inches to the gallon.

Six miles away, beyond the steamy banks of the Banana River, the million spectators who watch the titanic ship rise, shimmering in dreamlike silence, are suddenly hit by the sound leaping across the lagoons, a cataclysmic roar so intense that some go numb for a moment—a relentless shock wave beating the face and chest, rattling cars; wave upon wave of thunder louder and deeper than any thunder ever heard; and a crackling vibration that pierces the body again and again. Crude, fearsome, catastrophic, it is the sound of the biggest engine there ever was, a monstrous jackhammer that seems to shudder the entire planet. One holds one's eyes and ears, but it seems that even the skin can hear. The ship bores up through the blue sky like a comet and soon disappears, leaving only the laconic space talk on the loudspeakers to confirm the reality of a wondrously beautiful thing that has vanished.

Three years later Apollo 17, the last of the moon flights, rode a pillar of fire that turned the night sky orange-pink, a false dawn visible for five hundred miles. In the stands awaiting the launch that night, historian William Thompson saw lightning flash in the distance "like the boasting threats of a small boy backing away from a fight." Humans, he mused, "were turning the tables on the heavens and riding that comet out of earth." The shock waves from Apollo struck microbarographs at Columbia University with a force exceeded, in fact, only by the eruption of Krakatoa in 1883 and the impact of the great Siberian comet in 1908. "Man," wrote Norman Mailer, "now had something with which to speak to God."[2]

Searching the Media for Intelligent Life

Critics complained that President Kennedy was less concerned with God than with a small, round, beady-eyed man named Khrushchev and that the $24 billion car-rattling comet was but an alternate means to Russian humiliation, seized upon only after Kennedy learned that American science would be unable to desalinate the sea. Muckraking journalists relish such revelations, less from cynical realism than from an adolescent resentment of authority, a paranoid envy of power that sees all history as a political circus. Yet the moon shot could not succeed as a tactical ploy if the feat itself were without inherent meaning. In the end, Kennedy's politics explain only the timing, not the larger motive for going to the moon.

Others complained that the money spent on Apollo should have gone to health and welfare, a naïveté often bound to a parental universe where evil is not inherent in human nature, no one is faulted for his own fate, and the risk-free society eludes us only for lack of lowest-common-denominator education, fiat money, and committees of honest men. If we are denied a reasonable responsibility for our own entitlement, and if conflict and inequality are seen not as inevitable to the human condition but as symptoms submissive to legislation, then a collective hypochondria begins to consume our larger visions. Life degenerates to an immediacy in which means become ends in themselves, and the crowning purpose is an animal sense of well-being. It is a world without awe, diminishing and trivializing all that it touches.

The postured carping of the critics persisted even though the annual cost of Apollo was only 10 percent of that spent on health and welfare, 4 percent of the military outlay, and one-third of what Americans spent on alcohol or cosmetics. We bet more in a single year on horse races than on the entire Apollo project. But the distortion is no surprise in a custodial society where the mythic is so mired in cynical pragmatism that 15 percent of those polled believed the moon shots were faked. Perhaps the disenchantment is that of a nation entering middle age. "The youth," wrote Thoreau in his journal, "gets together his materials to build a bridge to the moon, or, perchance, a palace or temple on the earth, and, at length, the middle-aged man concludes to build a woodshed with them."

Yet the complaints about man on the moon were often no less inane than the apologetics, which ranged from prestige and power to Teflon pans, pocket calculators, and powdered orange juice. One puzzles over the disparity between the wonder of the feat and the meanness of the public perception. A partial answer is that much of the public is ill-represented by the media. In dank bars and around department-store TVs, where crowds fell silent to see a fellow man touch the moon, the deep sense of wonder was grist too fine for the mills of slicks and Sunday supplements. Yet the mood was short-lived. When later moon shots were allowed to interrupt football coverage the networks were deluged with angry phone calls. It was as though the first moon landing were another of those fads cataloged by *Women's Wear Daily* ("IN: Nehru jackets, carved Javanese monkeys, moonwalks"). The loss of interest was encouraged by promoting the moon landing not as a historic beginning but as a political termination. The result, as Apollo 11 astronaut Michael Collins observed, was to view the landing as another televised spectacular, a Super Bowl of technology. Why was the same Super Bowl being played over and over again? The *New York Times* reported Apollo 11 with the largest headline ever: MEN LAND ON THE MOON. By Apollo 16 the story had deteriorated to the *Washington Post* epitaph: TWO KLUTZES ON THE MOON.[3]

The Sacred Grove

The one image from the moonflights to survive public apathy was perhaps the one least anticipated. There is a moment in the documentary *For All Mankind* (1989) when the familiar footage of astronauts clowning aboard the moon-bound spacecraft ceases and the giant theater screen goes black. Then up from the lower left corner, like a lost balloon, drifts the bright Earth, marbled milky white on sapphire seas with patches of sunbaked brown, floating, fragile as a Christmas ornament, vivid on the black velvet of space. Rolling through the cosmic night like a child's ball, without apparent purpose yet holding all human meaning, it is "an extraordinary kind of sacred grove," said Joseph Campbell, "set apart for the rituals of life."[4]

When the astronauts stepped onto the barren, bombed-out surface of the moon ("magnificent desolation!" said Aldrin) they found that the only meaningful object was that wisp of color afloat in the black sky, four times larger and eight times brighter than the moon from Earth, containing all that we know and are ("O, a meaning! / over us on these silent beaches," wrote Archibald MacLeish).[5] Able to cover with a thumb the site of all human history, astronauts felt both solitary detachment and profound communion, for they saw no national borders, only the ancient seas and continental rafts, swathed in luminous clouds. Like the simple but wondrous gifts brought to primitives by European explorers, the astronauts had put a mirror in the hand of humanity.

The sight of "that tiny raft in the enormous empty night," as MacLeish described it, "that bright loveliness in the eternal cold," continues to transform our self-image in ways yet unknown.[6] The spread of the ecology movement in the early seventies reflected a new solicitude toward the planet, an ironic result of the attempt to escape her. The living Earth, floating like a space flower above the horizon of the dead moon, became a symbol not only for the unity of all life but for the reenchantment of the world, for new mythologies merging science and mysticism in a yearning for planetary consciousness. The mythic moon—Diana, Luna, Melusine—now branded with bootprints, dissected in laboratories, and littered with NASA's debris, was brought down to earth while the Earth was placed in the heavens and given the name Gaia. The pitted surface of the moon, glinting gunmetal gray in the sun, seemed to reflect the desolation of the modern ego, while the living Earth became the "round object" of depth psychology, the archetype of the larger Self—inner and outer, head and heart, unfragmented by abstract society.

The moon seemed a fitting destination for Newtonian man, sealed with his own waste in the urban capsules that had been his triumph, as distanced from the primal sources of meaning as the moonwalkers were from Earth. But like the city, the lunar landscape, as Neil Armstrong noted, has a "stark beauty all its own," a changeless wilderness where rolling, sunny slopes gleam like virgin snow and thousand-foot gorges border majestic, three-mile-high mountains—lifeless, windless, looming still and serene, only the harsh shadows moving with the sun. Yet the beauty of this permanence is that of the cemetery. Such well-named sites as the Sea of Tranquility and the Sea of Serenity reflect the sublimity of death itself.

The photograph of Earthrise encouraged the shift in focus from outer to inner space. It was an example of cultural eversion, in which a process taken to its extreme becomes its opposite. If the moon landing was the culmination of a half-millennium drive to gain dominion over nature, then the sight of our fragile, lonely world seemed pivotal in turning the quest inward—not just to psychology, mythology, and metaphysics, but to the new self-awareness that comes when one realizes that the parent is mortal. The photograph enabled some to see Earth as an organism capable of death. "As with a childhood home," said astronaut Jack Schmitt, "we see the Earth clearly only as we prepare to leave it."[7]

If the medieval view of an anthropocentric universe resembled the narcissistic perspective of childhood, then the Copernican image of a world adrift in an indifferent cosmos was in the spirit of adolescent alienation (and inflation, for accepting our diminishment made us greater than the gods). The harbinger of our maturity, then, may be the image of the whole Earth. For while it brought the first direct experience of the Copernican reality, it also suggested the emptiness of twentieth-century posturing about humanity's tragic absurdity and despair—an attitude rooted in the Calvinist contempt for man and its compensatory injunction to perform glorious works. For some, the logical end of that four-century imperative was the faceless astronaut. Adrift at the end of his umbilical, or alone on the lunar waste, he seemed a symbol of the modern paradox: isolation and impotence within a womb of wondrous works. But he also heralds a coming of age through the perfection of those works. Seeing our own essence in the floating Earth, we are finally more sacred than our works. "No longer that preposterous figure at the center, no longer that degraded and degrading victim off at the margins of reality," wrote MacLeish, "man may at last become himself." "We shall not cease from exploration," wrote T. S. Eliot, "And the end of all our exploring / Will be to arrive where we started / And know the place for the first time."[8]

Old Devil Moon

If putting the Earth into the heavens had universal appeal, putting the moon underfoot did not. The realities of spaceflight were in many ways disappointing to the very group who should have felt most fulfilled: the science-fiction fans. The treatment of interplanetary flight in this genre, which has always drawn a disproportionate number of fans in their teens and early twenties, reflected the power/innocence ambivalence so common to adolescence, with regressive fantasies of maternal utopias on one hand and visions of omnipotence on the other ("Ord sat in his swivel chair and surveyed the solar system"). If one escaped the castrating bonds of Mother Earth in great phallic rockets heroically bound for an unknown future, the ships were also fragile vessels, warm and womblike in the awful void, reaching for some eternal home. The appeal of interplanetary fiction had rested on the adolescent fantasy of transcending the parental world while remaining safely within it, on a romantic longing for the inaccessible, on dreams of heroic adventure, and on the hope that sentient beings—or at least some great secret—lay in the sunlit meadows of remote worlds. Most compelling was Mars, with its polar caps, apparent seasons, and penciled canals, a world abiding like the Sphinx in the desert of space, awaiting new life. Generations of boys had gazed at that peaceful point of light low in the evening sky, yearning to run barefoot on her soft red sands, to wander up a ridge and discover her dusky moons, rising on a rose-red city half as old as time.

How sobering, then, to see the moon invaded, between commercials, by inscrutable supermen who seemed inseparable from their machines, insulated from danger and adventure alike, unable to sift the silver dust in their fingers or feel the icy cold of another world. They had dropped, it seemed, onto a scarred wasteland of cinders and ash, not the silvery moon of song, the green-cheese moon, the moon that rhymed with June, croon, and spoon, moon of lunacy and harvest lovers, Carolina Moon, Old Devil Moon, swollen and orange, riding a purple evening. "If only they had found footprints, fishing line, the trace of horses," lamented one poet; another asked, "Dare we land upon a dream?"[9]

While the desacralized moon reinforced for many the new ecologic devotion to Earth, it sent the science-fiction fans to Mars. Unlike the Earth-first faction, however, they valued the moon landing as a triumphant if somewhat pedestrian plateau in humanity's pyramid to the stars.

The Great Pyramid

The space program stands with the cathedrals and pyramids as one of the great "central projects" of history, epic social feats embodying the worldview of a culture and the spirit of an age. On the launch pads, the rockets point heavenward like Gothic spires. Searchlights intersect on a waiting ship to form a great candescent pyramid, ablaze on the black horizon like some alien encounter, radiating light to the heavens. To reach for the heavens seems almost the signature of the central project. The pyramid was called the "stairway to heaven," the cathedral the "gate of heaven." The archetype is in Genesis: "let us build us a city and a tower, whose top may reach unto heaven."

Literature on the pyramids, cathedrals, and moon shots has tended to miss the significance not only of great height as the signal feature of central projects, but also their function as means through which whole cultures have found symbolic expression. Writers often pay lip service to the official rationales—immortality for the Pharaoh, a shrine to the Holy Virgin, or the quest for the grail of lunar rocks—while stressing the negative function of these projects as a source of shallow political pride ("let us make us a name," said the builders of Babel) or as a display of collective vitality ("And the Lord said . . . now nothing will be restrained from them which they have imagined to do"). Though some have noted that the central project focuses the energies and educates the consciousness of a population during periods of cultural transition, attracting the best and most adventurous minds of the age,[10] most of the interpretation has been narrowly political, reflecting the pedestrian, power-oriented, if not paranoid, slant of contemporary social science. Thus the pyramids become a ploy for political control, the Gothic cathedral is rooted in royal squabbles, and the space program is but a product of WASP prejudice or cold war hypocrisy—themes that lack all perception of the projects as spiritual quests in the broadest and deepest sense.

Rising from a sunbaked plateau on the west bank of the Nile, the Great Pyramid at Giza is a dark silhouette on the dusty sky, jutting off the desolate land like some inscrutable monolith from Martian lore. For fifty centuries the pyramids have stood silent, shimmering in the heat of the desert like ghost ships on the sea of time, looming stark and lonely against the twilight, sentinels from a lost civilization. Oldest of the Seven Wonders of the Ancient World, the Great Pyramid was built around 2560 BC from two and a half million blocks of limestone, some weighing up to seventy tons, with seams one-fiftieth the thickness of this page. A hundred thousand men toiled for twenty years without wheel, horse, or iron tools to complete the fourteen-acre, forty-story megalith, once flawless and gleaming white.

The monumental scale and minute precision of this enduring enigma seems so incongruous with the technology of its time that it has given rise to occult hypotheses—that the pyramids are products of divine intervention, that they were launching pads for alien spaceships, or that they embody arcane knowledge from the lost civilization of Atlantis. The dimensions, proportions, and siting of the Great Pyramid suggest that the builders had not only an advanced astronomy but a precise knowledge of pi, of the circumference of the Earth, and even of the flattening of the poles. But most prominent in the architecture is its embodiment of the Egyptian worldview—the sense that experience is ultimately to be understood subjectively, or heuristically, in terms of rhythm, harmony, proportion, symmetry, balance, and such numerological archetypes as paradoxic polarity, dialectic trinity, and equilibrated quaternity. The construction gangs who built the pyramids were not made up of slaves but of volunteers, for the means did not exist to control 50,000 to 150,000 men against their will. They seem even to have developed a competitive team spirit, and those who worked on the sacred project were honored in their villages. The Great Pyramid, in short, was more than a colossal headstone for a superstitious and narcissistic king, more even than a means for inaugurating the nation-state into history. It was the apotheosis of Egyptian science and culture. The building of the Great Pyramid, as Lewis Mumford has observed, spawned the first social megamachine composed of specialized human parts. Looming on the edge of the Libyan desert, this man-made mountain, almost half a mile square, was the Egyptian space program.[11]

The Gate of Heaven

Consecrated in 1260, the cathedral at Chartres rose sheer and silver-gray out of a wheat-rich prairie southwest of Paris. Armies of masons and glassmakers labored for generations to create what is now the oldest survivor of the Gothic space project. People came from all over France to drag stones from the quarry and pass bricks from hand to hand, chanting hymns as they labored and singing the holy songs around the night fires. Lords and ladies pulled carts with the rest and whole villages vacated to accommodate the laborers. The work was done in perfect discipline, often in profound silence. Flying buttresses multiplied in a dizzying play of arcs upon arcs; the north tower rose to 375 feet; and nearly 4000 figures were set into 176 huge windows. Inside, clusters of frail columns soared upward and flared out into the cross-ribbed vault eleven stories above the floor. Space seemed to grow larger as one looked from the dim lower area up into the almost insubstantial region of the windows, floating bursts of stained-glass color. A supernal violet light transformed the space, as though one were inside an immense flower that had grown naturally from the earth. "This is a place of awe," intoned the bishop when the church was finally consecrated. "Here is the court of God and the gate of Heaven."[12]

The early Gothic cathedral seemed literally to reach for the heavens: "a mighty hymn in stone and glass . . . flinging its passion against the sky."[13] The impression was intensified on the interior by strong vertical lines that made the building seem even higher than it was. Outside, the structure seemed to be drawn skyward by some relentless countergravitational force that raised and stretched every protuberance like cat's fur under an amber rod. The pointed Gothic arch, said one architect, "is a bow always tending to expand." "Hidden within its tensions," adds Loren Eiseley, "is the upward surge of the space rocket."[14]

More important than the Gothic effort to soar into the heavens (the actual height was comparable to that of the Great Pyramid and the Apollo vehicle) was its attempt to capture a sense of celestial space on Earth. Unlike the Romanesque, with its thick columns, heavy arches, massive wall space, and dark, gloomy interior, the Gothic is airy, delicate, and light. One suggests earthly power, the other spiritual transcendence. The pointed slits in the Romanesque wall are replaced by the great rose windows of the Gothic, made possible by the flying buttress, which allowed great spaces to be filled with stained glass and stone tracery, flooding the interior with light. The effect is that of a porous or transparent wall that light permeates and transfigures. The ethereal impression is strengthened by the fact that the light is subdued by the deep color of the windows. The result of all these soaring lines, airy spaces, and astral light is an aura in which matter itself seems insubstantial and weightless. The whole structure, down to the two-ton stones in the base of each pier, seems in a perpetual moment of liftoff.

This attempt to escape from Earth was more than symbolic. In the Gothic view the symbol itself was the reality; the physical world, a subjective product of man's limited perception, was only a means to the transcendent—the visible reflected the invisible, and the symbol was perceived not as illusion but as revelation. This applied above all to sacred architecture. The cathedral was seen not as a metaphor but as a model of the cosmos, just as scientific paradigms model the modern universe. With neither the empiricism nor the technology to explore the medieval belief that souls literally resided in the sky, simulation of the heavens on Earth became the goal of the Gothic space project. Perhaps the similar endeavor of science fiction to bring outer space into the great dark movie palaces of the fifties foreshadowed the rebirth of this vertical outlook—the contemporary spiritual resurgence that has followed the five-century interregnum of horizontal empiricism, centuries in which consciousness grew not inward and upward but outward in geographic and philosophic breadth.[15]

A Dream of the Child in Man

In spite of the horizontal drive of modern Western culture, pyramid-building has in a sense never ceased, persisting through ziggurats, temples, and cathedrals to the modern pyramid of science and technology—the rise, brick by empirical brick, of Newtonian know-how over the last half-millennium. From the tip of that pyramid Armstrong took his "one small step" onto the surface of the moon. Mankind takes its "giant leap" as a technological creature whose abstract shape is itself a contemporary pyramid of specialized expertise. It is this creature who reaches the moon, where the role of the astronaut—the cell at the apex of the pyramid—is virtually to bend and pick up stones. Though individuals may fail, the creature is unremitting, much as an army of Brazilian ants will bridge a moat with their own corpses to descend on a plantation house.

Loren Eiseley suggested that our species resembles a blight spreading on a fruit, a spore bearer, devouring the planet's resources. Microscopic slime molds swarm into concentrated aggregations, thrusting up spore palaces like city skyscrapers, the rupture of which may disseminate the living spores as far away proportionately as our journey to the moon. It is conceivable, wrote Eiseley, "that man's cities are only the ephemeral moment of his spawning—that he must descend upon the orchard of far worlds as a blight is lifted and driven, outward across the night," or die.[16] Or perhaps we steer spaceship Earth like Ahab on the *Pequod*, our means sane, our motive and object mad, drawn by the whiteness of the Milky Way into the starry deep, lashed to the object of our search with the lines of the chase.

Yet Ahab may be but a pathology in the process of matter coming slowly to know itself through humanity, the nervous and reproductive system of Gaia, longing for immortality among the far worlds. Just as we have seen the Earth from space, our progeny may gaze back on the radiant presence of the home galaxy, 600,000 trillion miles across, drifting in the cosmic ocean like a great whale, singing her inscrutable song to the radio telescopes of a hundred billion sister creatures. Perhaps some four-dimensional child of the cosmos will ultimately look back on the universe itself—a hundred-billion-trillion-mile spongelike mesh made of fine strands of galactic light, like a great glowing brain.

On the other hand, there are those who hold that only the mind may roam at will among the stars, for the moon is little more than a light-second away in a universe at least 15 billion light-years across. In this expanding and contracting "cosmic prison," as Eiseley calls it, "the diastole of some inconceivable being," our function and duration compare to that of a mayfly, and our tragic destiny to that of the classical hero, the hubristic flaw lying in our very ability to conceive of transcendence.

Yet in that irony may lie our salvation. In his series *Cosmos*, Carl Sagan is the lone voyager in the vaulted, cathedral-like interior of a ship that drifts majestically among stellar fires and mysterious worlds. His journey, we discover, is not outward but inward, home to the Earth where the ship began as a dandelion seed, floating from

Sagan's fingers on the sea breeze, borne up out of Earth to become the brilliant burst of light-spines that is the ship's external shape. Not a view of the far future, but a vision set from the beginning in the seed of evolution, it is the cathedral of the imagination. That it comes from the human core is underscored by the story of John Merrick, Victorian England's grotesquely deformed "elephant man." A gentle soul retrieved from a freak show and imprisoned in a hospital where he was often mistreated as an animal, Merrick died at twenty-seven of a ruptured spinal cord due to the weight of his own horrifying head. In his room he left a magnificently intricate model, built from scraps of cardboard trash, of the soaring spires of St. Philip's, the cathedral visible from the window of his own cosmic prison.

The ship of the mind, the cathedral of the imagination, has been the heart of the central project. One returns in the end to great height—a reaching for the heavens—as its essence. Each of the grand central projects has at its core a romantic idealism detached from all things practical and political. The great strength and beauty of the pyramids lay in the utter uselessness of the final product. The central project is the dream of the child in man—to raise a mountain in the desert, to build a celestial palace above medieval huts, to cavort on the sands of another world. Like all final concerns, it arises not from the ethic of work but from the spirit of *play*. Unlike the modern mistaking of means for ends, it is a taproot into human meaning. After all the apologetics, we will walk on Mars, not from false pride or foolish praxis, but in the spirit of wonder that sets our species apart.

The image of man on the moon that will endure is not the flag or the science but humans at play—the boyish exuberance, the pratfalls and belly flops, Gene Cernan bursting into song as he bounced like a rabbit with his basket of rocks down the Valley of Taurus Littrow, Al Shepard teeing off on Fra Mauro, Duke and Young yahooing as they bounded over the undulating plain in their moon jeep like lunar Keystone Kops. These are the images that will last, just as the toys will remain where we left them, the rover poised for its next task amid footprints chiseled in the dust, as fresh in a million years as if the driver had only stepped away for a cup of coffee.

But our dream is finally of a living cosmos, teeming like the silent world beneath the sparkle of the sea. It is the hope that one day, on a distant island world, we will *discover* an abandoned vehicle, encounter the footprint of a poor savage Friday or the soaring cathedral of some alien elephant man. It is the outward dream of the edge and the inward dream of the center, of touching the unknown depths of our larger selves, of going Home.

The Reenchantment of the World

The astronaut, sealed safely in his portable environment but unable to touch the world on which he walks, has become the symbol of postmodern man. We yearn to reconnect with nature, with one another, with ourselves. A fragmented, depersonalized, demythologized society spawns cancerlike individuals who lack a sense of anything larger than themselves and who thus destroy the social organism that sustains them. Simultaneously alienated and inflated, the ego becomes obsessed with order and control, barricading itself into a tiny clearing in the dark forest of the soul. Clinging to a false self, one loses receptivity to the symbolic unconscious, and with it the ability to experience wonder. The future of spaceflight is threatened less by the cynical journalist or the philistine politician than by the atrophied imagination, the withered capacity for wonder that afflicts the modern mind. Our sense of wonder may survive in fleeting visions of birth and death, origin and destiny, ocean and stars. But as snowflakes melt on the well-traveled walk or the morning star is lost in the light of day, dreams dry up in a drone of listless talk on bad help and great linebackers, echoing through empty fortresses, untouched by wonder. Drained from immediate experience, awe has been removed to the edge. The oceans and night sky are the modern preserves, the national parks of wonder, though our frenzied cities extinguish even the stars.

Yet events in recent decades have begun to reawaken an awareness of the unconscious, of the something-larger-than-the-ego that is the basis for all wonder. And if we are all astronauts afloat in the amniotic fluid of the abstract society, it is from that same womb that we may be born into the cosmos. In the last century we have had our own Copernican revolution. We have known for less than a lifetime that our galaxy is lost among a hundred billion other whorls of light, the fingerprints of creation, embedded in the "dark matter" that may compose 90 percent of the universe. And just as the early explorers sailed west in search of the East, the Newtonian science that rose in their wake now finds its own form of eversion. Its sturdy ships of reason, having explored to the far edges of rational experience, encounter the abyss that Columbus did not. On the macro- and microscopic frontiers, attempts to account for black holes or for particles that defy time and space begin to sound like the ancient mystics. As classic paradigms crumble, merging inner and outer unknowns, we again see the night sky as analogue to the soul; the old mysteries trickle back, the old enchantments return to the world.

The dream of spaceflight is in the end a search for roots, for something fixed and eternal. "Our identity," wrote Eiseley, "is a dream. We are process, not reality," prisoners of the concept of linear time, which creates the illusion of form and permanence.[17] The analytical psychologist would explain that the heavens, representing the outer unknown, are perceived as an analogue to the unconscious, the inner unknown; the longing for the stars is the quest for the greater Self, the equivalent of God.

Thus the idea that the moon, like Mallory's mountain, must be conquered "because it is there" is deceptive, for the moon is less a taunting adversary ("We've knocked the bastard off!" said Hillary atop Everest) than a piece of the infinite mystery passing nearby. Even the Promethean conquest of nature itself, the Apollonian thrust of the last half-millennium, cannot explain the fascination with space. The spectacular launch of Apollo was as misleading, in fact, as its name. Visions of fire ships and phallic rockets, of metal behemoths tearing man from Mother Earth, snorting flames as birds scatter in terror, are of secondary significance. The urge to enter space is rooted less in power than in innocence, less in agency than in communion. Though we may still seek to reenter Eden astride the machine, an eversion of emphasis in the latter half of this century has made the machine incidental to the postmodern imperative: to recontact the Great Mother, to reconnect with meaning.

Seeking the Center at the Edge

It would be wrong to see this quest for home as simply a longing for the lost mother. As with all things fundamental, the reality is a paradoxic tension; light and dark, yin and yang, inner and outer are finally inseparable. Like Columbus, we enter space seeking the East in the West, journeying, as Joseph Campbell said, "outward into ourselves."[18] If there is a common thread through all world mythology and religion, it is that the nature of man and the cosmos are one. It matters not whether this is literally true in some holographic sense, or subjectively true in that consciousness must constellate experience within its own limited spectrum; for us, it is true nevertheless. In the gleam in my wife's dark eye burns the Great Galaxy in Andromeda, while somewhere out on its myriad worlds recur the forgotten sound of my father's laugh and the scent of my mother's hair.

So we dream of sailing a cozy ketch out past a million suns, across the dark sea of the soul, in search of the mirror lake in the soft green meadow, the secret center, the glint in the eye of God. Even if it is not *in* time or *in* space, there is still a sense, in the cathedral of the mind, that the journey toward the horizon will bring it closer. "Come, my friends, 'tis not too late to seek a newer world," cried Tennyson's Ulysses, "to sail beyond the sunset, and the baths of all the western stars!" From the dawn of recorded history, the westward course of the sun has been that of rebirth and moral regeneration. "The ultimate effect of the discovery of the new world," wrote historian Charles Sanford, was "to substitute for the spiritual pilgrimage of Dante and Bunyan the 'way West' as the way of salvation."[19] From John Winthrop's "City upon a Hill," the Puritan moral example to the old world, down to our own nostalgia for the purity of the frontier, the way into the wilderness has been the way home. As with evolution itself, the way backward is lost—to the primordial sea, to the personal world of the primitive, to rambling twilight talk on a small-town porch, to our own elusive youth. As the Western horizon recedes across oceans and continents and out into the cosmos, the quest goes forward and outward, seeking the center at the edge.

In the VIP stands for the night launch of Apollo 17 sat a 124-year-old ex-slave named Charlie Smith.[20] Just as the astronaut became a sensing element for all humanity, this last survivor of the age of Jackson had traveled through time to be the eyes for an entire population long decaying in the earth, for those who had cleared the wilderness that became Houston. Gazing across the dark lagoon with their last witness, those pioneers might have wondered less at the thunderous fire ship than at the eerie apathy of their progeny. Ironically, the view of the Earth from space confirmed for many the notion that we should tend our own garden, rescue our own house, and make our pilgrimage not to the stars but to the far shores of our own kindred souls. Yet it is the unipolar mind of the Hebraic-Christian tradition that sees outer and inner space, like matter and spirit or pride and humility, as mutually exclusive. The reality is the paradoxic equilibrium of the two in tension. We *are* such polar tensions, like the note in a vibrato, or the rhythms and cycles that define all things. The whole person must have both the humility to nurture the Earth and the pride to go to Mars. Love and wonder, center and edge are finally the same, the solitary key to the prison of the self.

Where there is no vision,
the people perish.

—Proverbs 29:18

Chapter Four

Abandon in Place

A S APOLLO 13 REACHED THE BLACKOUT PHASE of reentry, flight director Gene Kranz was on his feet behind the row of consoles in Mission Control, pacing and smoking as he always did at critical moments. Kranz was tall, lanky, and big-boned, counted among the rarefied circle of "steely-eyed missile men" for his ability to issue cool, clipped orders while juggling a hundred problems competing for his attention. Besieged by men demanding emergency priority, each for his own piece of the whole, Kranz would listen quietly, ask a question or two, say, "Gentlemen, I thank you for your input," and issue a six-word decision in his staccato style. With a face like a wedge cut into rough planes, and a blond buzz cut that was barely visible in bad light, the in-tense, hawk-eyed Kranz had the look of a legendary drill sergeant. "It was like boot camp," he once said. "We took our best controllers and made them instructors. They hammered away at the weak links until they broke. If a guy had a drinking problem or a personal problem, he either forgot about it or he cracked. . . . Some guys died of heart attacks. A couple of them committed suicide."[1]

In the windowless "trench" at Mission Control, crises were handled in cryptic murmurs, and the internalized tension surfaced in nods, glances, and technical acronyms. The more calm things appeared, the worse they probably were. If the job of flight controller was intensely demanding, the role of flight director was superhuman. Orchestrating twenty minds, each focused on one small segment of the mission, Kranz was the real skipper of the spacecraft. He was responsible not only for encyclopedic knowledge of a machine with 15 million parts, but also for a massive flight plan, a vast body of mission rules, and a barrage of input from many simultaneous sources, often with only seconds to make life-and-death decisions—all under the gaze of the world. Kranz called it the "finest job in the spaceflight business." Like a maestro, he had not mastered every instrument yet he caught every mistake, and he knew how the whole must sound. From that acronymic dissonance came the concerto of haloed continents, the adagio of distant Earth, and the ballet of mankind on the surface of the moon.

The Eye of the Tiger

Apollo 13 was not the first crisis Kranz had met. He had seen the Mercury-Redstone launch its escape rocket while the silent booster threatened to explode. He was on duty when Neil Armstrong spun out of control during Gemini 8. And his quick thinking had rescued the lunar module during Apollo 5. "Gene was the guy you wanted on the headset when there was trouble," a fellow flight director once said. Thus the most dangerous part of the first moon landing, the descent to the lunar surface, was assigned to Kranz, who had to make "go" or "no go" decisions in the face of onboard computer failures, faltering telemetry, a field of boulders, and only seconds of remaining fuel ("You'd better remind them there ain't no damn gas stations on the moon").

But no crisis had matched that of Apollo 13. No flight-control team had ever been asked to make so many life-and-death decisions, improvise so many fixes, or sustain their nerve for so long a time. For four days following an oxygen tank explosion, snowballing crises threatened every aspect of survival—power, water, oxygen, fuel, guidance, communications, engines, reentry, and recovery. Yet what would seem a hopeless predicament was overcome moment to moment, point by point, as the crew scavenged and improvised, resorting finally to penciled math and seat-of-the-pants piloting. The need to shout transmissions, the hiss of equipment, and the icy cold of the near-dead spacecraft kept sleep to less than three hours a day. Water condensed in pearl-sized drops on the metal bulkheads, which seemed to draw heat from the body and dispel it to the stars. Two days after Jack Swigert spilled water on his shoes, his feet were still frozen. Fighting a kidney infection, Fred Haise shivered violently for four hours until Jim Lovell hugged his body to give warmth. Frogmen who opened the hatch on the tropical sea could still feel the chill of space.

The networks did not carry the crew's television broadcast, conducted only minutes before the explosion, for after the first lunar landing the nation had lost interest in moon shots. But with the news that lives might be lost in space, there was a sudden worldwide resurgence of concern. Apollo 13 came to symbolize both the nobility and fallibility of humanity—the heroic defiance that reaches out to the stars and the tragic hubris that lures us to the flame, only to plunge us into the abyss. Thus the gaping gash running half the length of the command module drew inevitable allusions to the *Titanic*, that mythic memorial to Edwardian hubris. But the real root of the association lay in the vision of being swallowed into a vast, mysterious deep, to lie forever in some cold, dark, unearthly place.

That Apollo 13 escaped *Titanic*'s fate was due to the intense composure of both the crew and the team in Houston. For Kranz and his band of controllers, that failed mission was ironically their finest hour—or 141 hours, during which Kranz survived on catnaps, remaining at the complex around the clock. His first job, after admonishing everyone to "keep cool" and not make things worse by guessing, was to preserve as many options as he could while closing down others. If he couldn't achieve initial goals, could he achieve lesser ones? "Eventually," he has said, "those goals may come down to mere survival, but even then you don't let yourself doubt that you can accomplish them. . . . If everybody worries, 'Well, if this thing happens or that thing happens,' you're not going to solve anything. Our job is to make things happen."[2]

When it did come down to mere survival, it was Kranz who made the critical decision to let the moon's gravity sling the spacecraft back toward Earth rather than risk the attempt to turn around and power home. He moved his White Team (the eight-hour shifts had color names) to a windowless chamber full of conference tables where they stayed for three days without leaving—looking, as he put it, "into the eye of the tiger." In addition to surmounting what seemed an impossible shortage of power and consumables, the assignment of his renamed Tiger Team was to completely rewrite the reentry checklist—normally a three-month project—in three days.

Emerging successfully, the White Team was back at the helm for the crucial reentry, Kranz pacing and smoking, wearing the one eccentricity he allowed himself: the symbolic white vest made anew for each flight by his wife, Marta. (All the more prominent against his narrow black tie and no-nonsense visage, the vests had become rallying banners for his White Team.) The four-minute blackout phase of reentry, when friction temperatures create an ionization cloud around the ship that blocks all communication, was especially tense given the possibility that the explosion had damaged the heat shields. As the four-minute mark passed without response from the crew, the only sounds in the room were those of the air conditioners and the hum of electrical equipment.

"All right, Capcom," said Kranz, grinding out the cigarette he had lit four minutes ago, "advise the crew we're standing by."

"*Odyssey*, Houston standing by, over," Joe Kerwin called. Nothing but static came back from the spacecraft. After two more attempts, over five minutes had passed with nothing but noise on the communications loop. The men at the consoles stared fixedly at their screens. The guests in the VIP gallery looked at one another.

Then Swigert's voice came through the crackle. Kerwin closed his eyes and drew a long breath. Kranz pumped a fist in the air. As the huge chutes appeared on the main viewing screen, Kerwin had to shout over the roar of cheers in the control room: "*Odyssey*, Houston. We show you on the mains." Unable to hear the response, he added, "Got you on television, babe!"[3]

One recalls the end of *The Bridges at Toko-Ri* when the admiral, gazing from the carrier as the surviving planes return, asks with quiet wonder, "Where do we *get* such men?"

A More Heroic Time

Born in the Ohio heartland at the height of the Great Depression, Eugene Francis Kranz grew up in an America whose watchwords were cooperation, persistence, discipline, and endurance. Losing his father at the age of six on the eve of the war, he came of age in years of personal and national hardship. From the world of Dunkirk and Iwo Jima, where risk and sacrifice defeated one of the darkest demons in history and left America a colossus astride the Earth, Gene Kranz learned the value of wholehearted communal engagement, seeing in crises not an excuse for self-pity or wishful fantasy, but an opportunity for creativity, for pursuing the possible in the present, for expressing the human spirit.

Earning a degree in aeronautical engineering from St. Louis University in only three years, Kranz flew F-86 fighters in Korea and was later a flight-test engineer. He entered the space program at its inception in 1960 and became one of the first flight directors for the Gemini missions. A devout Catholic, unabashed patriot, father of six, and fervent believer in the exploration of space, his abiding attachments were to God, country, family, and his band of brothers in Flight Operations. Kranz was a man who could get misty at the sound of "The Star-Spangled Banner," and he played Sousa marches in his office to jump-start his day. But it was a documentary on Apollo 13, made almost a quarter-century after the event, that revealed a sentimentality deeper than his Sousa tapes.

At the end of the two-hour program, as the main chutes deploy to soulful music, Kranz, whose flat Midwestern voice had provided much of the technical commentary, is shown in close-up as he remembers the jubilant mood in Mission Control—the cheering, the lighting of cigars, the waving of small American flags. Then, pausing in midsentence, shaking his head, Kranz wells up. Struggling mightily—militarily—to maintain his composure, he presses his lower lip hard against the upper, then manages a near-whisper: "It was neat."

The night the documentary aired, it was seen in a hotel room in Los Angeles by the team that was planning the film *Apollo 13*. What affected them most was Kranz's lapse. "If we've got a story that can make Gene Kranz cry," one of them said, "we've got a story."[4]

But why tell a story whose ending everyone knew? Those who went to the film with that question found that the story was less about spaceflight than about the human spirit. And Kranz, who emerged no less heroic than the astronauts, was invited on the lecture circuit. He came one evening to the old auditorium on the Stanford campus, heavier for his years but having lost neither the crew cut nor his intense commitment. To most of that student audience, born a decade after Apollo, he might as well have been Lindbergh remembering his Ryan monoplane, or Edison recounting the first "Hellohhh" shouted onto a wax cylinder. Like the computer, spaceflight has become commonplace, albeit endless circling in near-Earth orbit. But in the postmodern void, there is a visceral need for heroes. And for these children of the custodial society, coming of age amid a failure of nerve, the film had struck a chord, a latent sense of human potential, a faint echo of a more heroic time, now as remote as the moon itself.

The story Kranz told revealed a more vital past—a time of common cause, intense teamwork, and transcendent purpose. The challenges, he said, had been met by young people, often no older than the students in the audience, people with a proud vision, with dedication, perseverance, and self-discipline. It would have been easy to see a Boy Scout squareness about this military man, passing out his "Mission Operations" sheet with its symbolic emblems and bold-print words—"confidence," "toughness," "teamwork," "responsibility." Yet those young people of more timid times, who had come out in sheets of rain to hear the man from the movie, rose to their feet—as though for a whole generation— and applauded at inordinate length.

The Dream of Animal Comfort

It has been more than a quarter-century since the last man left the moon. Remnants of the great rockets now lie in the Smithsonian with the *John Bull*, the Tin Lizzie, and the *Spirit of St.*

Louis. On the abandoned launch pad at the cape, dry grass bends in the sea breeze. In another quarter-century it is likely that all twelve who walked on the moon will have passed into history. And there are now two generations who cannot remember when spaceflight was still a dream, who view Armstrong's leap as an archival event, another Lindbergh commotion.

Even in the days of Apollo the public lost interest after the first lunar landing. Having beaten the Russians in the Super Bowl of space, we went back to business and Monday Night Football. After six walks on the moon, the last three were scrapped, saving seven-tenths of 1 percent of Apollo's total cost. Yet the $24 billion Apollo program cost each American only a dollar a month for nine years, and this ignores exponential returns to the economy. The $38 billion spent on space between 1961 and 1972 was barely 1 percent of the national budget, 3 percent of the amount allotted to social programs, and half the figure for detected welfare fraud.[5] Had even this moderate commitment to space persisted, we would have walked on Mars a decade ago.

Rationales for the abrupt ending of Apollo suggested that the scientific potential did not warrant the cost, that the program stole technical talent from other fields, that it had little military value, and that Apollo's dinosaur boosters were a dead end in the evolution of spaceflight, which must now consolidate a stepping-stone presence in Earth orbit. Historian Walter McDougall argues that the space effort was part of an ideological package that Americans purchased after *Sputnik* in the belief that the United States must adopt the technocratic model to get back on top. By the early seventies, with the relaxation of cold war tensions and the growing concern over domestic issues, "the original model for civilian technocracy, the space program, became dispensable."[6] This may well have been true, but it does not seem to justify scrapping three paid-up moon shots. It is far more likely that President Nixon watched Apollo's TV ratings drop and decided that the risks—especially after the near disaster of Apollo 13—outweighed diminishing returns.

A part of the problem was television itself. Ghostly images of astronauts on the moon revealed little detail, while the image of Earth in the lunar sky was a blurred white fleck segmented by two or three picture lines. The limited audiovisual spectrum of television could not communicate even the earthbound wonder felt by those present at the launch, though the event involved the crude forms and simple contrasts (a large, isolated, unidirectional object spewing fire) most suited to television. And beyond the technical limits of the media lay the confines of mass consciousness, restricting spaceflight coverage to hardware, technical routine, cost-benefit ratios, or the lifestyles of the astronauts, trivia that soon became boring. Like the iconic cars that dominate our ads and movies, journalism in a pragmatic, means-become-ends culture is expected to *get* us somewhere. Television, moreover, becomes the great leveler of experience. How far away was the moon? The same distance as Vietnam—across the family room.

In a larger sense, the shift from the world of print to the world of television—from the discourse of ideas to the surrealism of objects without context—has drastically shortened our attention span and corroded our ability to create images. The 6000 hours of television that the average American child has seen by the fifth year floods the young brain with images at the very time it would otherwise learn to generate them from within, a condition unconducive to the growth of imagination. Ironically, the means through which we watched the moon walks had itself reduced our ability to see their significance. We are, as Neil Postman notes, "amusing ourselves to death."[7]

But it is not the mass media that are finally to blame for public apathy toward manned spaceflight. If we lack the imagination to infuse the event with wonder, the fault lies in ourselves. A vast number of us are simply uncurious about anything we cannot perceive directly (making Mars even less interesting than the moon). Since the universe of modern science, with neither center nor edge, violates the archetypes we call common sense, many choose simply to ignore it. The alleged 15 percent who believed that the moon shot was an elaborate government hoax staged for television exemplified in the extreme the widespread want of the most elementary concepts necessary to grasp the event. When the 1994 Los Angeles earthquake knocked out the power in the middle of the night, Griffith Observatory received numerous calls asking whether the quake was responsible for the sky being "so weird." The citybound callers had never seen a star-filled sky.

No less profane were the critics who bemoaned Apollo's lack of utility, calling it escapism, military adventurism, fodder for technocracy, or a triumph of the WASPs, views often rooted in paranoid distrust of power and authority.[8] For the same reasons that polls name the president of the United States the greatest living American year after year, the average citizen fears and reveres power above all else, perceiving the world as narrowly political and measuring most things by their instrumental utility.

Without exception, polls have shown that interest in space rises proportionally with education and income. In forty national polls measuring public support for the space program between 1965 and 1994, an average of only 18 percent of the respondents favored its expansion when given the choice of funding other programs. In recent polls, only 8 percent would increase expenditures for space exploration, only 9 percent feel informed on the subject, and only 22 percent express any real interest in it.[9] The great majority of Americans have put space at or near the top of the list of proposed budgetary cuts, favoring the reduction of programs perceived as having little effect on their daily lives.

When columnist Drew Pearson complained, in the wake of Apollo, that Nixon had chosen to fund "the most unnecessary project of the century" rather than to clean "240 million gallons of excrement" out of the Potomac, the *National Review* noted that "out of such stuff as sewage treatment plants are liberal dreams spun."[10] Having risen with the urban-industrial middle class, liberalism has suffered from its own success. The belief in equal opportunity has become a protest against unequal results. What began as a reaction against artificial pockets of wealth culminated in the denial that character and intelligence need correlate with well-being. And the bold social conscience that once stood against the evils of industrialization became a pathological crusade to neutralize every conceivable stroke of ill-fortune. Confusing people with livestock—creatures to be doctored and fed—the liberal agenda has degenerated to a quest for animal comfort.

The Geography of the Soul

In the long run, the whole politics of society is more profoundly changed by a new sense of human potential than by any amount of obsessive self-maintenance. "Where there is no vision," says the proverb of Solomon, "the people perish." Without a source of meaning larger than the ego or beyond mere survival, one is left at the center of a universe devoid of transcendence. The significance of anything derives from its larger context, one dependent in turn on still greater perspectives, until we reach what sociologist Peter Berger calls the "sacred canopy," the boundary where known and unknown meet. This largest conceptual cosmos halts the infinite regression of "Why" questions, closing the perceived order with such symbols as "God" or "Universe," just as mathematical order is preserved by capping the system with notions of zero and infinity. These ultimate constructs need not belong to organized religions, whose ossified idolatries are often inimical to larger meaning. Any living symbol of the boundary, left inchoate and mysterious, becomes an object of awe and wonder, veiling some great mystery of indeterminate size and origin.

Wonder, in its larger sense, denotes the *mysterium tremendum*, the aura of unfathomable majesty, utterly humbling and wholly Other, surrounding the sublime and terrifying unknowns that border our models of reality—the dark forest, the empty desert, the sacred mountain, the boundless sea, the black silence of cosmic infinity. Thus we gaze into the night sky and feel not diminishment but dilation. We sense the vastness and passion of creation and glimpse an equally vast interior—the "enormous geography of the soul," as journalist Edwin Dubb put it.[11] We are aware of the stars only because we have evolved a corresponding inner space.

Thus, the depth to which we experience the cosmos is proportional to our level of self-awareness. And with a capacity for self-transcendence and connection to larger contexts of meaning and identity, the ego perceives its own finitude, free of false self-images and able to see others as more than means to its own unexamined agenda. Like the astronaut looking back on an Earth without political boundaries, one sees the multifaceted surface and oceanic depths of the larger Self; one reflects on reflecting, a boundless regress that projects a deeper interiority onto the outer world, which becomes a more profound field of wonder.

At the opposite extreme of self-transcendence is self-assertion—an inflated ego that subsumes its social context much as a cancer cell proliferates to

no purpose and metastasizes. Psychologist Abraham Maslow argues that the excessively self-
assertive are governed by such lower needs as security and acceptance, measuring others by their
usefulness, mistaking wealth and power for ends
in themselves. Spirituality is confined at best to
conventional creeds, idolatries that avoid the
larger questions. A false self projects its own system of categories and expectations onto the
world, recognizing only those things that can be
quickly labeled and filed away in some well-worn
category. Extremes of boredom are avoided by a
rapid succession of stock perceptions, as with spec-
tator sports or TV fare, where the elements are re-
arranged to effect suspense or surprise without
deepening the experience. Thus, while everything
seems familiar, little is really seen or known.

Embedded in the protective cocoon of the culture much as the infant is unable to distinguish
between self and other, the unreflexive, self-
assertive person substitutes know-how for know-
why, finding security in the familiar to the point
of feeling adrift without a fixed pattern or routine. This condition can develop when an over-
anxious parent or repressive culture leads a child
to distrust the world, instilling a feeling of impo-

tence and fear of novelty that curbs exploratory play. Conversely, a child forced into independence too early may develop obsessions with power and mastery. At either extreme, the consequent fear of the unknown, of facing internal uncertainty or losing external control, can bring a compulsion toward order and predictability, and a reductionist dismissal of anything unrelated to the practical, empirical world as irrelevant nonsense—symptoms that roughly describe the pathology of power-oriented politicians who obstruct the exploration of space. Anxiety kills curiosity and the urge to explore, while the want of self-reflection strangles imagination, creativity, poetry, and the sense of wonder.

In contrast, the self-transcendent personality is motivated by growth needs rather than deficiency needs, seeing an experience as one would a sunset or a work of art, not as a means to unexamined ends but as an end in itself. Fascinated by the frontiers of the familiar world, such people break through the cultural cocoon to the ineffable, inexhaustible Other. They are directed not outward at what they lack but inward toward expressing what is intrinsic to the organism. Maslow has argued that the "metaneeds" for truth, goodness, beauty, order, and unity are biologically as basic as the lower needs but remain obscured and elusive for those whose growth was arrested at earlier levels.[12]

In addition to the fulfillment of primary needs, however, self-transcendence requires the reflexive capacity to see beyond the raw content of experience. In a phenomenological study of introversion, psychologists Joel Shapiro and Irving Alexander distinguish between the "lived" experience of the extravert, which is direct and unmediated, like a cropped photograph, and the "reflective" experience of the introvert, which is more like a painting. In extreme form, the extravert's self is simply equivalent to the path of his encounters as he moves through the world, whereas the introvert's self is an evolving inner construct in light of which raw experience is selectively interpreted and incorporated. This incessant and intensive creation of meaning can leave the introvert hypersensitive and susceptible to overload, while the extravert is more stable and thus better armed for the arenas of wealth and power. The explicit meaning that results from the introvert's self-reflection contrasts with the implicit, "felt-meanings" of the extreme extravert, who is bored by any experience that must be re-flectively transformed to gain significance.[13]

Most of us, of course, lie nearer the middle. But to the degree that we lack self-awareness we can-not think metaphorically and are thus con-demned to the literal life. Arthur Koestler contrasts the "routine," which operates on a sin-gle linear plane, with the "creative," which imaginatively associates two previously uncon-nected matrices of thought. The juxtaposition of matrices disarms unreflective self-assertion (as absorption in the car chase is displaced to the theater when the film slips out of focus), dislodg-ing the subjective experience from its objective correlate and revealing the inner world as a sep-arate reality.[14] It is the essence of the metaphoric imagination, analogous to the dreamer's aware-ness of dreaming—what Blake called "double vi-sion," the ability to simultaneously perceive a thing in at least two ways. A recent study of ninety-one prominent creative people found that their personalities contain a complex balance of polar extremes—cooperation and aggression, re-alism and idealism, rebellion and tradition, masculine and feminine—a stereoscopic depth that retains a childlike delight in the strange and the unknown, a sense that anything mysterious, regardless of utility, is worthy of attention.[15]

Psychologist Heinz Kohut notes that such people seem to preserve the child's capacity to experience reality with less of the "buffering ego" that protects the average adult from traumatization, but also from creativity and discovery.[16] The "buffer" seems to be rational consciousness itself. To the degree that we rely on deliberate conscious thinking—our analytic, reductionist, problem-solving mind—we lose access to the slower, less conscious ways of knowing that are the seedbeds of creativity. A simple example is the fact that we often recall a name only when we cease deliberately trying. Cognitive scientists now suspect that the function of rational consciousness is simply to invent plausible rationales for decisions already made at preconscious levels, its evolutionary role being to analyze and evaluate these inclinations in threatening situations.[17] In modern competitive society, however, the threats are more often to status or self-esteem. Pressured by such uncertainties, and fearing that we will lose control or commit condemning errors, we revert to rigid, clear-cut, and conventional thinking, adopting one shallow nostrum, one fashionable idea after another.

We find ourselves in this benumbed state when rational consciousness eclipses the larger unconscious self—that part of the mind, grounded in feeling, that gives meaning to experience. The linear, analytic mode also reduces the multivalent and paradoxic symbols of the deeper mind to mere conceptual signs ("Mother Nature," retaining the Latin root, becomes "matter"). Unlike the convergent thinking of focused consciousness, the undermind is divergent, intuitive, associational, leisurely, playful, and tolerant of ambiguity. In touch with this pre- or semiconscious realm, we are more self-aware, less threatened by the unfamiliar, and more willing to explore without knowing the object of our search. Psychologist Guy Claxton argues that chronic stress has caused modern culture to lose sight of this mode, to view it as a threat to reason and control, a "wild and unruly 'thing' that lives in the dangerous Freudian dungeon of the mind." The result is a withered capacity for creativity, curiosity, and wonder. The root condition is often an authoritarian environment—overprotective parents who impose rigid models of "right" thinking and feeling on children who comply for fear of abandonment, or traffic-cop teachers with inflexible rules and standardized routines, operating in a system geared to the lowest common denominator. Both produce either shallow rebels or cooperative conformists, neither of whom are capable of self-transcendence.[18]

The idea that stress and the need for control blocks access to the creative undermind is supported by the finding that children burdened with early responsibility tend to develop "thick boundaries" and lack imagination. One study speculates that the brain in such cases may achieve the permanent synaptic pathways of adulthood earlier, while the paths of the more creative, "thin-boundaried" people remain closer to the formative state of childhood, less definite and specific, with more branching and ambiguity, like wandering back roads versus an all-purpose freeway system.[19] More obvious is the fact that excessive demands, whether rooted in early responsibility or cultural brainwashing, require rigidity, reducing the time and motive for probing one's experience; and the less one explores, the less comfortable or compelling exploring becomes.

A synthesis of these various perspectives may lie in recent insights into brain lateralization. It has become clear that the left side of the brain deciphers the *text* of experience while the right side provides the larger meaning or *context*. This qualifies in subtle ways the older distinction between temporal/linear/analytic and spatial/analogic/synthetic. The difference is more one of foreground versus background, deduction versus induction, or disconnected particulars versus a coherent, overall understanding of an event. People with right hemisphere damage seem to lack a larger sense of what is going on within the self and between the self and the world. The literal nature of the left hemisphere cannot attribute two meanings to one thing; it does not experience Koestler's intersecting matrices. Unable to think metaphorically, such people are slow to grasp humor, sarcasm, or irony; and they lack the reflexivity necessary to a sense of wonder.

No less vital than access to the right hemisphere is hemispheric differentiation itself. It seems likely that the lateralization of the two hemispheres (like two eyes or two ears) has a stereoscopic effect. The wider the differentiation the greater the ability to hold a feeling or concept in one mind while looking it over with the other—a prerequisite to creativity and a sense of wonder. Studies have shown, in fact, that creative people have high interhemispheric communication. One suspects that the transcendent perspective requires a balance of polarized hemispheres—ideally, an equilibrium of the two minds.[20]

In the end, it is balance that is vital in all these polarities. To tread the thin line between rational consciousness and the creative undermind, between instrumental knowledge and multivalent meaning, is to avoid both rigid dogma and paralyzing indecision. In a balanced society, merging the "two cultures," the impetus for science and exploration would be spiritual in the deepest sense. Science would no longer languish in our schools and the will to explore would survive the Philistine pits of power.[21]

A Sense of the Cosmos

The point of all these models is that they delineate a spectrum of self-awareness, a measure of wonder that divides the apostles from the opponents of space exploration. Without this reflexive awareness the stars remain unremarkable; the night sky becomes peripheral.[22] But the models find larger meaning in the fact that the psychic development of each individual reenacts the evolution of human consciousness. For the balancing of self-assertion and self-transcendence defines the contemporary coming of age, a cultural transformation for which the quintessential symbol is the dream of spaceflight.

The development of the Western mind has had as its metanarrative the collective emergence of conscious ego out of preconscious union with nature, a schism that has involved an adolescent obsession with self-assertive power that is archetypally masculine. From the patriarchal religions and rationalist philosophies of the Hebraic-Greek tradition to the objectivist science of the modern era, the Western canons have driven toward atomistic societies of autonomous, existentially free individuals.

This evolution, however, was founded on a repression of the feminine, a progressive denial, as historian Richard Tarnas notes, of the *anima mundi*, of the community of being, "of mystery and ambiguity, of imagination, emotion, instinct, body, nature, woman." The point is less that Western history is a chronicle of chauvinist imperialism than that it has been a preordained stage of growth. The driving impulse of the West's masculine consciousness has been its "dialectical quest not only to realize itself, to forge its own autonomy, but also, finally, to recover its connection with the whole," to differentiate itself from the feminine but then rediscover and reunite with it at a higher level of consciousness.[23]

At the core of the current cultural transformation—beyond regressions to primitive spiritual traditions—is a collective reunion with the larger Self at a higher turn of the spiral. The greater our reflexivity, the closer we come to this waking dialogue of conscious and unconscious. Coming of age, we cease our adolescent cycling between poles of the human condition—isolation and communion, power and innocence, masculine and feminine. We approach the point at which the increasingly isolated ego finally accepts its own finitude, aware that it is a small island on a great dark sea. Projected outward, it is what Jacob Needleman called a "sense of the cosmos."[24]

Explore or Expire

Are we the spores of spaceflower Earth, the metamorphosis of Gaia to Galaxia, or are we a planetary cancer, metastasizing to the biosphere and near planets? Any living system is but one level in a hierarchy that ascends through cell, organ, organism, family, community, society, species, and ecosystem (matter or spirit), the meaning and identity of each level lying with its function in the larger whole. Each intermediate structure displays a tension of self-assertion and integration, behaving as a self-contained whole in relation to subordinate levels and as a dependent part in relation to the larger reality. While self-assertion can become a metastasis, destroying the greater whole, the result of self-transcendence is evolution.

Living systems reach out to their environment, merging with larger systems in the fight against entropy. We know from the new science of chaos and complexity that when an open system interacts with the environment, it resists the slow accumulation of stresses until a breaking point is reached, just as water suddenly boils or an oppressed people revolt. While the energy of a closed system is continually dissipated, moving it toward a maximum disorder of random particles, living systems maintain themselves by importing energy from the environment, processing it, and discarding waste. The evolutionary result is a self-organizing synthesis toward ever more complex structures whose goal is maximum order. As long as the fluctuations caused by the continuous flow of energy are minor, the system damps them. But if the fluctuations reach a critical size, they "perturb" the system—the elements of the old pattern come into contact with one another in new ways. The parts reorganize into a new whole, and the system may escape into a higher order.[25]

It is at its frontiers that a species experiences the most perturbing stress. The urge to explore, the quest of the part for the whole, has been the primary force in evolution since the first water creatures began to reconnoiter the land. We humans see this impulse as the drive to self-transcendence, the unfolding of self-awareness. The need to see the larger reality—from the mountaintop, the moon, or the Archimedean points of science—is the basic imperative of consciousness, the specialty of our species. If we insist that the human quest await the healing of every sore on the body politic, we condemn ourselves to stagnation. Living systems cannot remain static; they evolve or decline. They explore or expire. The inner experience of this imperative is curiosity and awe. The sense of wonder—the need to find our place in the whole—is not only the genesis of personal growth but the very mechanism of evolution, driving us to become more than we are. Exploration, evolution, and self-transcendence are but different perspectives on the same process.

Three billion years ago, as tidal rhythms exposed the creatures of the shallows to sun and air, it was the moon that summoned life from the sea. Now the moon calls again. Either we have a destiny that transcends the individual or we will ultimately succumb. And though the creative vision may elude the average person, "the highest of his duties," wrote the biologist J.B.S. Haldane, "is to assist those who are creating, and the worst of his sins is to hinder them."[26] The least among us, reflected Walter Cronkite after the moon landing, are improved by the feats of the best of us.

Fooling Us Out of Our Limits

Those who compare the legacy of the Renaissance with the promise of space argue that the ingrown homogenization of Europe in the sixteenth century now applies to the whole planet, to which space offers not only rich new veins of empirical knowledge but a deprovincialization of the spirit. Space may save modernism from the black hole of solipsism. For if literary golden ages coincide with peaks of frontier expansion, spawning a Homer, Shakespeare, Melville, Conrad, or Twain, the closing begets Tennessee Williams and Jack Kerouac. "We had to explore into outer space," wrote Mailer, as the "last way to discover the metaphysical pits of that world of technique which choked the pores of modern consciousness—yes, we might have to go out into space until the mystery of new discovery would force us to regard the world once again as poets, behold it as savages who knew that if the universe was a lock, its key was metaphor rather than measure."[27]

The frontier, like the world of the child, is a place of wonder explored in the act of play. Work is self-maintenance; play is self-transcendence, probing the larger context, seeking the higher order. And like the sacred spaces reserved for sport—cool green fields hidden in the bowels of Pittsburgh and Chicago—the frontier is seen as hallowed ground, where the gray monotonies of mortal limits are forgotten and each moment is eternal.

Joseph Campbell has observed that in countless myths from all parts of the world the quest for fire occurred not because anyone knew what the practical uses of fire would be, but because it was fascinating. Those same myths credit the capture of fire with setting man apart from the beasts, for it was the earliest sign of that willingness to pursue fascination at great risk that has been the signature of our species. Man requires these fascinations, said the poet Robinson Jeffers, as "visions that fool him out of his limits."

Like the capture of fire, the longing for spaceflight is rooted less in means than in meaning itself. Beyond all the pragmatic apologetics, there is a certain unlikelihood about the Egyptian pyramid, the Gothic cathedral, or the Saturn 5 rocket; a towering, unearthly presence on the Libyan desert, the Florida coast, or the wheatlands of southern France. Like all final concerns, these central projects arose not from the ethic of work but in the spirit of play, their great strength and beauty lying in their utter impracticality. It is difficult enough for Americans, whose values and traditions are rooted so deeply in the work ethic, to conceive of an imprudent project at the center of any life, let alone at the core of an entire culture. Yet the Protestant ethic itself may be the greatest central project of all. Ironically, the pursuit of means as ends in themselves—equating wealth with happiness, power with success, isolation with freedom, change with progress—spawned the technological pyramid that has freed the mass of humanity from mere utility.

One result has been to obscure the boundary between art and knowledge. In primitive times the two were identical; with the modern notion of objective science, art became subjective. But in the postmodern era, figure and ground are reversed; knowledge itself becomes the work of art. "In a world in which men write thousands of books and one million scientific papers a year," says historian William Thompson, "the mythic *bricoleur* is the man who *plays* with all that information and hears a music inside the noise."[28]

Perhaps, says Daniel Boorstin, we are no longer merely *Homo sapiens* but rather *Homo ludens,* "at play in the fields of the stars." Perhaps we have learned to luxuriate in the expanding universe of expanding mysteries, "where achievements are measured not in finality of answers, but in fertility of questions."[29] It is not that we go into space seeking a literal edge. For even more revolutionary than the shift from an Earth- to a sun-centered cosmos is the modern universe, which lacks any center at all. Rather, we stand alone on the leading edge of evolution, exploring our horizons as children probe the world in play.

Lured by the hope that life is no less characteristic of the cosmic ocean than of the terrestrial, it is humans who must go into space, to wander far worlds and meet once more the dread unknowns, the dry-mouthed fears of the old explorers. The galaxies, great continents in the cosmic sea, arrayed "in knots and streamers across billions of light-years, like motes of dust dancing in window light," may lie eternally beyond touch.[30] But the *telos* is the endless quest itself— matter expressing itself as spirit, spirit finding its epiphany in matter. To believe less—or to believe more—is to live in the shallows of what it means to be human.

Were some catastrophe to destroy the human species, alien explorers might one day discover on the Florida coast a great stone table, standing Sphinxlike against a gray blend of sea and sky. Stenciled with the words ABANDON IN PLACE, it was intended to stand forever. Perhaps they would see it with other enigmatic monoliths—Stonehenge or the pyramids—as having some religious function. They would be correct. Abandoned since the days of Apollo, Pad 34 rests like an ancient ruin at the center of a great circle, the old roads radiating in all directions. Lofty arches formed by the four legs of the great launch table frame another arch in the distance; and beyond that, lying in the scrubby weeds, the giant flame deflectors gather rust. There is something eternal about the scene, something elegiac in the lonely arches, crumbling at the edges, abiding in silence but for the song of dry grass in the sea breeze, requiem not only for the three who died there in the fire, but for the greatest project in human history. Risen and fallen on a crest of idealism for which "Camelot" was no misnomer,

Apollo fades into the past with aging moon-walkers, with Gene Kranz and his Sousa marches—anthems for a Homeric age.

As monument to Apollo, Pad 34 merits a better inscription. Perhaps the final word belongs to Wernher von Braun, the man most responsible for Apollo's success. With his boundless energy and infectious enthusiasm, the tall man with the broad chin and the boyish smile embodied not only the triumph of Apollo but its fate as well. Ignored by NASA when he was no longer needed, he died of cancer at age 65. A young space enthusiast once asked von Braun what it takes to send a man to the moon. His answer is his epitaph: "The will to do it."

Along a parabola life like a rocket flies,
Mainly in darkness, now and then on a
rainbow.

Andrei Voznesensky,
Parabolic Ballad

Chapter Five

Reflexions

I WAS FIVE YEARS OLD when my father was shipped overseas to take part in the allied invasion of France. My mother and I, joined by his mother, had been following him from camp to camp through California and Texas, living in bug-ridden hotels. The irony is that while my father trudged through France, fearing he would not survive the war, it was my mother who died of lupus in the hallway of an overcrowded hospital in sunny Pasadena. Afraid to tell me the truth, my grandmother took me back to San Francisco with the assurance that my mother would eventually join us. As the weeks wore on, I was left to piece things together for myself. I have a searing memory of the night, by a window in our small apartment, when it all surfaced in my mind. It was like the dawning of consciousness itself. My memory of everything before that moment floats in the haze of childhood, a series of scattered images, but everything after that night lies in the larger world of adults, a coherent, structured chronology that seems to have followed a sudden leap in self-awareness.

It was then that I began to notice the night sky. For both death and the stars are symbols of the barrier, each a sacred canopy, one in space, the other in time; one outer, one inner. The notion of my nonexistence is as alien to my daily reality as the thought of a hundred billion galaxies, the finitude of self as unthinkable as cosmic infinity; and a true sense of either lasts but an instant. Without the mystery of death we would be unmoved by the mystery of the universe. Were there no threat of nonexistence there would be no sacred canopy, no projection of final meaning onto symbols of immortality. The two sides of wonder—the ecstasy of boundless possibility (*mysterium*) and the horror of absolute limitation (*tremendum*)—form the tension that spawns all art and science. But the ground of both is the awareness of death. From the refusal to accept its finality arose the pyramids, the cathedrals, and the Apollo rockets.

The awareness of death seems not only the price of consciousness but its measure as well. In the cul-de-sac of postmodern culture, the reality of death is repressed to the point of being the new pornography, viewed with the lurid fascination once reserved for sex. The denial of death defaults to competitive obsessions with power or to conventional doctrines of immortality that only further eviscerate curiosity and wonder. But like my night by the window, the mass traumas of the last century seem to have encouraged a collective leap in reflexive awareness. Deep within the malignancy of modern individualism is a longing to restore some larger context, some meaningful end for runaway means.

The search for what psychologist Robert Lifton calls the "symbols of immortality" intensifies as accelerating change undermines all the institutions—family, church, government, education—that have traditionally provided the larger context for those symbols. Thus it becomes ever more difficult to connect with something larger than the ego, to find a continuity with past and future that transcends time and mortality. A sense of immortality may come through our children, our contributions to posterity, an immersion in nature, or moments of ecstasy found in music, dance, sex, creativity, athletic effort, or intense camaraderie. Such experiences fill the need for transcendence, for an expansion of one's life space, while reducing death anxiety. But the deeper the awareness of death the more profound the moments of ecstasy.[1]

The word ecstasy means "to stand outside of," to be outside of oneself. To gaze at the stellar ocean is to step outside of linear time, merging the personal and transpersonal. As the star-filled night stretches away into space, the frozen panorama of my past reaches back over time, a single seamless event in which old conflicts fade and larger meanings come clear. Enlarging me in space and time, the stars become personal, my past universal.

One such experience, a literal step outside, remains as fresh in my mind as it was forty years ago. I was at a debutante ball in a sprawling pavilion on the edge of a lake. My strategy at such functions was usually to find a magazine in the lounge and await the redeeming ride home. In this instance, however, I was trapped at a table with other sixteen-year-olds vying with one another over whose family had the worst help and who, at one time or another, had been the most drunk. So when the dancing began I stepped outside into the night, wandering along the shore among the dark trees to the opposite side of the small bay.

With the lights of the great white pavilion reflecting on the lake, and the music floating over the water, the tiny figures dancing in the distance seemed fleeting as mayflies. Out beyond the bay, where the open lake stretched away to a dark line of distant forest, the Milky Way rose out of the ancient trees, vaulting across an ocean of stars. It was a moment akin to my encounter with the Pacific or my night in the mountain camp, but inner and outer were now in some way inseparable. The stillness of my past, the permanence of things remembered, seemed somehow hidden in those countless specks, burning a thousand lifetimes distant. And I knew that somewhere out on the faintest fleck, itself a galaxy of a hundred billion suns, lay a shimmering lake and the echo of music immersed in the moment—and perhaps, on some dark shore, one who had stopped to step outside, to gaze back across a billion years of space-time.

Perhaps in some recess of the psyche the human organism knows that it is a fractal of the reflexive universe itself, repeated in microcosmic multiplicity like a holographic plate—worlds within worlds within worlds—and that, far beyond the cold war reflex, our emergence into space rose inexorably from the taproot of evolution, the heartsong of a cosmos that is finally the dance of spirit. We are fascinated by the roiling surf for the same reason we are transfixed by fire: we too are matter asserting itself as energy. So are the fires of the night sky, where our being was written long before planets were born or oceans condensed or mortal cells emerged from primordial soup. The communion of cells formed a lung, a heart—an *eye*. And the world awoke. The price of vision was mortality, but its prize was the capacity for love and wonder. And if the cosmos is spirit incarnate, then the flame of life, the eye of consciousness, is its resurrection. Like salmon, we hurl ourselves against entropy, returning in fits and starts and occasional heroic leaps to our place of origin; as though the primal spirit had fallen into infinite multiformity and had somehow forgotten itself in the process. We are that ineffable essence, slowly, agonizingly, *remembering*.

A Candle in the Dark

It had been a dark and bitter year. The war languished in Vietnam, students rioted around the globe, the Soviets invaded Czechoslovakia, North Korea seized the USS *Pueblo*, a B-52 crashed carrying four hydrogen bombs, Chicago police battered demonstrators at the Democratic convention, Robert Kennedy was assassinated in Los Angeles, and Martin Luther King was shot down in Memphis. Discontent was epidemic, disillusion profound, as American families sat down to dinner on Christmas Eve, 1968.

Yet my most indelible memory of that evening is the hush of kitchen clatter as our gathering was drawn to the TV—children, grandmothers, cousins, in-laws, and old-maid aunts—to gaze through a spacecraft window at the mountains and craters of the moon, a phosphorescent world creeping across the screen, curving away to the black of space. "In the beginning," intoned a metallic voice across a quarter-million miles, "God created the heaven and the Earth . . . "; and the poignant closing of those first men to circle the moon: "Merry Christmas, and God bless all of you—all of you on the good Earth."

And once again the Earth seemed good. *Time* magazine scrapped plans to feature "the Dissenter" as Man of the Year, substituting the three astronauts. Like a latter-day star in the east, Apollo 8 had risen over a world longing for epiphany. It seemed proper that the event occur at Christmas—the last living myth in a disenchanted world, archaic as Genesis yet compelling at the core.

In the twenty-first century, the exploration of space will revitalize humanity just as Apollo 8 redeemed a dark year. In 1997, when *Pathfinder* put the first rover on Mars and the incoming pictures were relayed to a large screen in the Pasadena Convention Center, thousands of people of every age and color cheered each new image. Under the Martian panorama, among strollers and wheelchairs, backpacks and business suits, a profound sense of communion filled the huge room. Strangers conversed like old friends; for beyond the politicians and engineers this was *their* triumph, an ascension of humanity itself. *We* were on Mars!

"From the moment the first flint was flaked," to borrow Auden's phrase, space was fated to be the final canvas for expressing in bold strokes the inexhaustible soul of humanity. We are alive at the dawn of a new Renaissance, a moment much like the morning of the modern age, when most of the globe lay deep in mystery and tall masts pierced the skies of burgeoning ports, luring those of imagination to seek their own destiny, to challenge the very foundations of man and nature, heaven and Earth. Like the sailing ships that incarnated the aura of the Renaissance, or the great steam locomotives that embodied the building of America, the spacecraft is an emblem of the human spirit, probing the cosmos like a "candle in the dark."

The phrase belongs to a man who passed into history just before Christmas, 1996. More than anyone of his century, Carl Sagan reignited the sense of wonder in a world increasingly content to simply exist. Wonder was the core motif in the complex fugue of Sagan's life: the six-year-old at the World Exposition, awestruck by the utopian sights; the boy standing with outstretched arms in an open field, imploring the magic force that had carried Burroughs's hero to that blood-red beacon burning low in the night sky; and the proud ship *Voyager*, with its pictures of man and its heartfelt hellos from the people of Earth. Who but Carl Sagan would cast humanity's bottle into the cosmic ocean?

It was his rare gift to walk that razor's edge between romance and reality. He was the dreamer and the doer, the theorist and the activist, combining lofty speculations with cold, hard logic, balancing soaring wonder with unrelenting skepticism. A common theme running through his many-faceted career was his confrontation with what he called our failure of nerve—the self-indulgence that has taken this nation from the world's largest producer and creditor to the world's largest consumer and debtor.

While the media cycles between pseudoscientific solipsism and existential despair, offering epiphanies in the form of aliens come to eviscerate cows and rape rural housewives, the epiphany for Sagan lay not in a cosmos that comes across light-years to doodle in our wheat fields, but one to which *we* must make the pilgrimage, across an infinite regression of Archimedean points, sailing outward into ourselves. "Space *exploration*," Sagan insisted, is not endless circles in low orbit, tending weightless tomatoes, it is "*going to other worlds*." The continued exploration of the solar system, he argued, is a challenge that can bind together nations, inspire youth, advance science, and ultimately end our confinement to one vulnerable world.[2]

But Sagan's grand vision was of voyages on a stellar ocean teeming with life. His last gift to us, the film *Contact*, was criticized for depicting the first alien encounter in an overly sentimental manner. But the quest for intelligent life in the universe has never been less than a search for a reflection of the personal self, something we normally find only in another human being—someone who rescues us from solitude, who is our only enduring connection between "in here" and "out there," our only real communion with something larger than the ego. Humanity's longing for a place in the heavens is that same need writ large.

Envoi

Carl Sagan's memorial is that silent streak of light that arced out over the dark Atlantic one hot August night, bearing, at his behest, greetings from Earth in fifty-nine languages, music from many cultures, and pictures of life on a blue planet. Launched in 1977, *Voyager* would explore what may be the homelands of our descendants, returning breathtaking images of the outer planets before passing out of the solar system. Traveling a million miles a day, *Voyager* will leave the Oort cloud, the trillion or more comets that orbit the sun, in 20,000 years. After hundreds of centuries, it will cross the line where the sun can no longer hold an object in orbit and will enter the open sea of interstellar space, the great dark between the stars—to sail forever, as Sagan said, "through the starry archipelagoes of the vast Milky Way Galaxy."

It seems appropriate that the last voice on the *Voyager* recording is that of Sagan's five-year-old son: "Hello from the children of planet Earth!" For we are a species still in childhood, only now becoming aware of the true immensity and complexity of the cosmos. Carrying the hopes of humanity, the dreams of millennia, *Voyager* reaches out to life in a universe turbulent and mysterious beyond anything imagined by our forebears. And if odds of entering another solar system are very small, perhaps eternity is time enough.

Voyager was inevitable from the first gleam in the eye of the hunter-gatherer, from the first fire, wheel, and furrow; it was latent in the stirrup and the longship, in the creak of every caravel, the ring of every railroad spike, the lonesome howl of every lumber-camp harmonica. *Voyager* is the distillation of our essence, our offering in the cosmic cathedral, the voices of millennia echoed in the vault of night.

Generations will come and go, civilizations will rise and fall, and long after Earth is vaporized by the sun and humanity is either extinct or evolved into other beings, *Voyager* will drift silently onward, carrying the message through countless eons: that there was, at our time and place in the cosmos, an *awareness* that knew something of its world and something of itself, an imperfect people of irrepressible spirit, of mathematics and music, of love and wonder, who dared to dream of reaching the stars.

Space Chronology

Significant Firsts and Events
Mentioned in Text

1619	*Harmonice mundi* by Johannes Kepler.
1634	*Somnium seu Astronomia Lunari* by Johannes Kepler.
1865	*De la Terre à la Lune* by Jules Verne.
1877	Martian *canali* reported by Giovanni Schiaparelli.
1895	*Mars* by Percival Lowell.
1898	*The War of the Worlds* by H. G. Wells.
1903	*Investigation of World Spaces by Reactive Vehicles* by Konstantin Tsiolkovsky.
1912	"Under the Moons of Mars" by Edgar Rice Burroughs serialized in *All-Story Magazine*.
1919	*A Method of Reaching Extreme Altitudes* by Robert Goddard.
1923	*Die Rakete zu den Planetenraumen* by Hermann Oberth.

1926

3/16   First liquid-fuel rocket launched by Robert
        Goddard.

1929   *Frau im Mond*, film produced and directed by Fritz Lang.

1938   John W. Campbell becomes editor of *Astounding Science Fiction*.

10/31   Orson Welles's *War of the Worlds* broadcast.

1942

10/3   First successful A-4 rocket launched at Peenemünde.

1944

5/29   Chesley Bonestell's paintings published in *Life* magazine.

1946

4/16   V-2 launched at White Sands for U.S. Army by Wernher von Braun.

1947

2/8   Robert Heinlein's "The Green Hills of Earth" published in the *Saturday Evening Post*.

1949   *The Conquest of Space* by Willy Ley, with illustrations by Bonestell.

2/24   WAC Corporal sounding rocket mounted on a V-2 reaches space.

1950

6/27   *Destination Moon*, film produced by George Pal, directed by Irving Pichel.

1952

3/22   First spaceflight article in *Collier's* series by Von Braun and others.

1957

10/4   First artificial satellite, Soviet *Sputnik 1*.

1958

1/31   First U.S. satellite launched, *Explorer 1*.

7/29   NASA founded.

1959

1/2   First craft to leave Earth's gravity, Soviet *Luna 1*.

9/12   Launch of first Earth probe to impact the Moon,
        Soviet *Luna 2*.

1961

3/25   President Kennedy announces the goal of landing
        a man on the moon.

4/12   First human spaceflight: Yuri Gargarin, *Vostok 1*.

5/5   First American in space: Alan Shepard, *Mercury 3*.

1962

2/20   First American in orbit: John Glenn, *Mercury 6*.

8/14   *Mariner 2* launched, first successful flyby of any
        planet (Venus).

1963

6/16   First woman in space, Valentina Tereshkova in
        Vostok 6.

1964

11/28   First probe to Mars launched, *Mariner 4*.

1965

3/18   First space walk: Alexei Leonov, *Voskhod 2*.

6/3   First American space walk: Ed White, *Gemini 4*.

1966

1/31   Launch of Soviet *Luna 9*, first soft landing on the
        Moon.

3/1   First human artifact to reach surface of another
        planet, Soviet *Venera 3* impacts Venus.

3/16   First docking: Neil Armstrong and David Scott,
        Gemini 8.

3/31   Launch of first Moon orbiter, Soviet *Luna 10*.

1967

1/27   Gus Grissom, Ed White, and Roger Chaffee killed
        in fire on Pad 34, Apollo 1.

1968

4/3   *2001: A Space Odyssey*, film produced and directed
        by Stanley Kubrick.

12/24   Frank Borman, Jim Lovell, and Dave Anders orbit
        the moon, Apollo 8.

1969

7/20   First lunar landing: Neil Armstrong and Buzz
        Aldrin, Apollo 11.

11/14   Second lunar landing: Alan Bean and Charles
        Conrad, Apollo 12.

1970

4/13   Apollo 13 explosion.

12/15   First soft landing on Venus, Soviet *Venera 7*.

1971

1/31   Third lunar landing: Alan Shepard and Edgar
        Mitchell, Apollo 14.

4/19   First space station, Soviet *Salyut 1*.

7/26   Fourth lunar landing: David Scott and James Irwin, Apollo 15.

11/13   *Mariner 9* orbits Mars.

1972

4/16   Fifth lunar landing: John Young and Charles Duke, Apollo 16.

12/7   Sixth lunar landing: Gene Cernan and Harrison Schmitt, Apollo 17.

1973

3/29   First Mercury flyby, *Mariner 10*.

5/14   First U.S. space station, *Skylab 1*.

12/3   First Jupiter flyby, *Pioneer 10*.

1976

7/20   *Viking 1* lands on Mars.

9/3   *Viking 2* lands on Mars.

1977

8/20   *Voyager 2* launched.

9/5   *Voyager 1* launched.

1979   *Cosmos*, Carl Sagan's thirteen-part series, airs on PBS.

1980

11/12   *Voyager 1* arrives at Saturn.

1981

4/12   First shuttle flight, *Columbia*.

8/25   *Voyager 2* arrives at Saturn.

1983

6/13   *Pioneer 10* leaves the Solar System.

6/18   First American woman in space, Sally Ride.

1984

2/7   First untethered spacewalk, Bruce McCandless.

1986

1/24   *Voyager 2* arrives at Uranus.

1/28   *Challenger* disaster.

2/20   Space station *Mir* launched.

1989

8/25   *Voyager 2* arrives at Neptune.

1993

12/4   Hubble space telescope repaired in orbit.

1996

12/20   Death of Carl Sagan.

1997

7/4   Mars *Pathfinder* rover lands on Mars.

Notes

Preface

1. Norman Mailer, *Of a Fire on the Moon* (New York: New American Library, 1971), p. 37.

Chapter One

1. Copernicus's book, filled with internal contradictions and unsolved mysteries, and thus neither simpler nor more accurate than the Ptolomaic system, attracted interest only in academic circles, where it was generally treated as no less heuristic than Ptolomy's model. And "it was Kepler, first of all, not Galileo," notes his definitive biographer, "who freed astronomy from the bonds of Aristotelian physics" (Max Caspar, *Kepler* [New York: Dover Publications, 1993], p. 136).

2. Arthur Koestler, *The Watershed: A Biography of Johannes Kepler* (Lanham, Md.: University Press of America, 1960), p. 195.

3. Ibid., p. 251. Many modern concepts, of course, were not so much discovered as rediscovered; Plutarch, for example, contended that the moon was a small Earth inhabited by intelligent beings; the "music of the spheres" originated with Cicero.

4. See Marjorie Hope Nicolson, *Voyages to the Moon* (New York: Macmillan, 1960).

5. On seventeenth-century England, see E. M. W. Tillyard, *The Elizabethan World Picture* (New York: Vintage Books, n.d.); Wallace Notestein, *The English People on the Eve of Colonization, 1603–1630* (New York: Harper & Row, 1954); and P. B. Medawar, *The Hope of Progress* (London: Methuen & Co., 1972), p. 110.

6. See Charles L. Sanford, *The Quest for Paradise: Europe and the American Moral Imagination* (Urbana, Ill.: University of Illinois Press, 1961).

7. Koestler, pp. 36, 66, 224. On Kepler, see also Fernand Hallyn, *The Poetic Structure of the World: Copernicus and Kepler* (New York: Zone Books, 1990); and Wolfgang Pauli, "The Influence of Archetypal Ideas on the Scientific Theories of Kepler," in Bollingen Foundation, ed., *The Interpretation of Nature and the Psyche* (New York: Pantheon Books, 1955), pp. 149–240.

8. Robert Heinlein's story, "Universe," was first published in *Astounding Science Fiction,* Ma, 1941. It was expanded as Robert Heinlein, *Orphans of the Sky* (New York: Putnam, 1964).

9. Not only was Verne the first to place a trip to the moon on a mathematical basis, but he was also the first to associate rockets with spaceflight, equipping his artillery shell—

the first spacecraft to resemble the modern image—with small rockets for steering. With uncanny prescience, the Frenchman fired his ship from the tip of Florida, splashing down in the Pacific within three miles of the recovery site for Apollo 8, the first flight to circle the moon.

10. On the American search for order, see Robert H. Wiebe, *The Search for Order, 1877–1920* (New York: Hill and Wang, 1967). For insights into the psychology of science fiction and the nature of SF fandom, see William Sims Bainbridge, *The Spaceflight/Revolution: A Sociological Study* (Malabar, Fla.: Robert E. Krieger, 1983), ch. 7; Harry Warner, Jr., *All Our Yesterdays: An Informal History of Science Fiction Fandom in the Forties* (Chicago: Advent, 1969); Brian W. Aldiss and Harry Harrison, eds., *Hell's Cartographers: Some Personal Histories of Science Fiction Writers* (London: Future, 1976); and Sam Moskowitz, *Seekers of Tomorrow: Masters of Modern Science Fiction* (Cleveland: World Publishing Co., 1966).

11. William Graves Hoyt, *Lowell and Mars* (Tucson, Ariz.: University of Arizona Press, 1976), p. 15.

12. Carl Sagan, *Cosmos* (New York: Random House, 1980), p. 110.

13. Ray Bradbury, et al., *Mars and the Mind of Man* (New York: Harper & Row, 1973), p. 35.

14. Milton Lehman, *Robert H. Goddard: Pioneer of Space Research* (New York: Da Capo Press, 1988), pp. 226, 52.

15. Ibid., p. 23.

16. Walter Dornberger, *V 2* (London: Hurst & Blackett, 1954), p. 137.

17. The distinction is James Hillman's (*Revisioning Psychology* [New York: Harper & Row, 1975]). The merging of the two represents the Jungian goal of "individuation," or psychic wholeness, in which ego-consciousness becomes aware of the total psyche, the larger Self. Interestingly, Edward Edinger (*Ego and Archetype: Individuation and the Religious Function of the Psyche* [Baltimore: Penguin Books, 1973]) chose to represent the emergence of ego-consciousness as a small circle rising out of a larger one until its center lies entirely outside, where the ego awakens to the Self much as a beached fish might become aware of the ocean. The resulting diagram resembles a small satellite orbiting its mother planet.

Chapter Two

1. Willy Ley and Chesley Bonestell, *The Conquest of Space* (New York: Viking Press, 1949).

2. Although Robert Lippert had received story lines for *Rocketship X-M* prior to the contracting of *Destination Moon* and managed to release the film a month ahead of *DM*, his decision to proceed with it was nevertheless a direct result of *DM*'s extensive prerelease publicity. See Bill Warren, *Keep Watching the Skies: American Science Fiction Movies of the Fifties, Volume I, 1950–1957* (Jefferson, N.C.: McFarland, 1982), pp. 11, 13.

3. Michael Collins, *Carrying the Fire: An Astronaut's Journeys* (New York: Bantam Books, 1983), p. 253.

4. Leo J. Moser, *The Technology Trap: Survival in a Man-Made Environment* (Chicago: Nelson-Hall, 1979), pp. 48–50.

5. S. G. Schwartz, "Amour De Voyage," *Michigan Quarterly Review* 18 (Spring 1979): 266.

6. National Aeronautics and Space Administration, *Why Man Explores* (Washington, D.C.: U.S. Government Printing Office, 1976), pp. 18–20.

7. Loren Eiseley, *The Unexpected Universe* (San Diego: Harvest Books, 1969), pp. 67–73.

Chapter Three

1. Carl Sagan, *Cosmos* (New York: Random House, 1980), p. 5.

2. William Irwin Thompson, *Passages About Earth: An Exploration of the New Planetary Culture* (New York: Harper & Row, 1974), pp. 1, 3. Norman Mailer, *Of a Fire on the Moon* (New York: New American Library, 1971), p. 93. The microbaragraphs (reacting to Apollo 4) are noted by Richard Lewis in *Appointment on the Moon* (New York: Ballantine Books, 1968), p. 420. The "comet" was a huge meteorite.

3. Michael Collins, *Carrying the Fire: An Astronaut's Journey* (New York: Bantam, 1983), p. 474.

4. Joseph Campbell, *Myths to Live By* (New York: Bantam, 1973), p. 245.

5. Columbia Broadcasting System, Inc. CBS News, *10:56:20 PM EDT, 7/20/69: The historic conquest of the moon as reported to the American people by CBS News over the CBS Television Network* (New York: CBS, 1970), p. 165. This is the original version of MacLeish's poem, "Voyage to the Moon," read over CBS during the Apollo 11 coverage.

6. MacLeish's *New York Times* essay on Apollo 8, quoted in John Noble Wilford, *We Reach the Moon* (New York: Bantam, 1969), p. 206.

7. Harrison Schmitt, "The New Ocean of Space," *Sky and Telescope* 64 (October 1982): 327.

8. MacLeish quoted in Wilford, *We Reach the Moon*, p. 206. T. S. Eliot, "Little Gidding," in *Four Quartets* (New York: Harcourt Brace Jovanovich, 1971), p. 59.

9. S. G. Schwartz, "Amour De Voyage," *Michigan Quarterly Review* 18 (Spring 1979): 266. May Swenson quoted in Ronald Weber, *Seeing Earth: Literary Responses to Space Exploration* (Athens, Ohio: Ohio University Press, 1985), p. 82.

10. Willis Harman and Howard Rheingold, *Higher Creativity: Liberating the Unconscious for Breakthrough Insights* (Los Angeles: Jeremy P. Tarcher, 1984), pp. 174–175.

11. On the pyramids, see Lewis Mumford, *The Myth of the Machine: Technics and Human Development* (New York: Harcourt, Brace & World, 1967), chs. 1, 9; Kurt Mendelssohn, *The Riddle of the Pyramids* (London: Thames & Hudson, 1974); and John Anthony West, *Serpent in the Sky: The High Wisdom of Ancient Egypt* (New York: Julian Press, 1987).

12. Kenneth Clark, *Civilisation: A Personal View* (New York: Harper & Row, 1969), pp. 50–60; Harman and Rheingold, *Higher Creativity*, p. 169.

13. Kenneth MacLeish, "Legacy from the Age of Faith: Chartres," *National Geographic* 136 (December 1969): 866, 880. The second phrase was taken from Henry Adams.

14. Loren Eiseley, *The Invisible Pyramid* (New York: Charles Scribner's Sons, 1970), p. 84.

15. On the cathedral, see Otto von Simson, *The Gothic Cathedral: Origins of Gothic Architecture and the Medieval Concept of Order,* 3rd ed. [Bollingen Series XLVIII] (Princeton, N.J.: Princeton University Press, 1988), especially the introduction and part 1.

16. Eiseley, *Invisible Pyramid*, pp. 53–54.

17. Loren Eiseley, *The Unexpected Universe* (San Diego: Harvest Books, 1969), p. 76.

18. Campbell, *Myths to Live By*, p. 246.

19. Charles L. Sanford, "An American Pilgrim's Progress," *American Quarterly* 4 (Winter 1955): 302.

20. On Charlie Smith, see Grover Lewis, "A Conversation with the Nation's Oldest Citizen," *Rolling Stone*, February 1, 1973, pp. 22–26; Andrew Chaikin, *A Man on the Moon: The Voyages of the Apollo Astronauts* (New York: Viking Press, 1994), pp. 495, 498, 501; David Baker, *The History of Manned Space Flight* (New York: Crown, 1982), pp. 438–439; and "The Talk of the Town," *The New Yorker*, December 30, 1972, p. 22.

Chapter Four

1. Mike Gray, *Angle of Attack: Harrison Storms and the Race to the Moon* (New York: Penguin, 1994), p. 280.

2. Jeffrey Kluger, *The Apollo Adventure: The Making of the Apollo Space Program and the Movie* Apollo 13 (New York: Pocket Books, 1995), p. 189.

3. The account is taken from Jim Lovell and Jeffrey Kluger, *Lost Moon: The Perilous Voyage of Apollo 13* (Boston: Houghton Mifflin, 1994), pp. 332–333.

4. Kluger, *Apollo Adventure*, pp. 192–193. The description of the documentary is taken in part from Kluger.

5. See "Overlooked Space Program Benefits," *Aviation Week*, March 15, 1971, p. 11; Larry Geis and Fabrice Florin, eds., *Moving into Space: The Myths and Realities of Extraterrestrial Space* (New York: Harper & Row, 1980), p. 25; Jeffrey M. Elliot, ed., *The Future of the Space Program/Large Corporations and Society: Discussions with 22 Science Fiction Writers* (San Bernardino, Calif.: Borgo Press, 1981), p. 6; and "Apollo's Moon Mission: Here Are the Results," *U.S. News and World Report*, August 4, 1969, p. 27.

6. Walter McDougall, *The Heavens and the Earth: A Political History of the Space Age* (New York: Basic Books, 1985), p. 422.

7. Neil Postman, *Amusing Ourselves to Death: Public Discourse in the Age of Show Business* (Harmondsworth, England: Penguin, 1986).

8. In addition to Norman Mailer's *Of a Fire on the Moon* (New York: New American Library, 1971), see, for example, William D. Atwill, *Fire and Power: The American Space Program as Postmodern Narrative* (Athens, Ga.: University of Georgia Press, 1994); David Lavery, *Late for the Sky: The Mentality of the Space Age* (Carbondale, Ill.: Southern Illinois University Press, 1992); Dale Carter, *The Final Frontier: The Rise and Fall of the American Rocket State* (London: Verso, 1988); William L. Crum, *Lunar Lunacy and Other Commentaries* (Philadelphia: Dorrance & Co., 1965); and Amitai Etzioni, *The Moon-Doggle: Domestic and International Implications of the Space Race* (Garden City, N.Y.: Doubleday, 1964).

From whence comes this need to lay low all human triumphs by linking them to the inevitable ills of society and culture? Yes, all things are interconnected, but does that reduce the meaning of the Statue of Liberty to a lure for cheap labor? It is ironic that the critics of spaceflight accuse its celebrants of ignoring the complexities of political and economic reality; for the same critics seem ignorant of the greater complexities of human nature—the ability to simultaneously experience the same reality in many ways. The linear, analytic, overintellectualized perception of the world, for all its intricacies, fails to plumb the depths of true complexity.

9. Vincent Kiernan, "Study Finds Space Support Dwindling," *Space News*, February 27–March 5, 1995, p. 6; 1993 National Opinion Research Center survey cited in "The Moon Landing Revisited," *American Enterprise,* July–August 1994, p. 89. A poll commissioned by Rockwell International in 1992 found that only 2 percent of the respondents thought they were extremely familiar with space activities (Paul S. Hardersen, *The Case for Space: Who Benefits from Explorations of the Last Frontier?* [Shrewsbury, Mass.: ATL Press, 1997], p. 165). In a poll taken just after the *Pathfinder* landing on Mars, 13 percent felt the government was not spending enough on space, but they were not asked to choose between space and other programs (Humphrey Taylor, "The Harris Poll on Space: Stronger Public Support but not for Spending More," *Space Times* 36 [November–December 1997], 15).

10. "Flat-Earth Liberals," *National Review*, July 29, 1969, p. 738.

11. Edwin Dubb, "Without Earth There Is No Heaven: The Cosmos Is Not a Physicist's Equation," *Harper's* 289 (February 1995): 40.

12. See Abraham H. Maslow, *Toward a Psychology of Being*, 2nd ed. (New York: Van Nostrand, 1968); *The Farther Reaches of Human Nature* (New York: Viking Press, 1971); and *Religions, Values, and Peak-Experiences* (New York: Viking Press, 1970). A similar polarity is suggested by Ernest G. Schachtel in *Metamorphosis: On the Development of Affect, Perception, Attention, and Memory* (New York: Da Capo Press, 1984), substituting autocentric/allocentric for self-assertive/self-transcendent. The autocentric experience is one of socially shared autism, a private world of the many in which one perceives only the expected conventional schemata. Without the rigidity of such familiar labels and signposts, the autocentric faces the terrifying abyss of the unknown and unmanageable. The knee-jerk, means-become-ends orientation of such embeddedness compares to the instinctive level of lower mammals, who react to any change with avoidance. Higher mammals, with an emerging sense of self-other separation, have developed the drive to explore, albeit one vulnerable to arrest.

13. Kenneth J. Shapiro and Irving E. Alexander, *The Experience of Introversion* (Durham, N.C.: Duke University Press, 1975). The conventional view of the introvert as withdrawn is, of course, far too simple. There is some basis, however, for associating paranoia with the extravert, who needs to control the path of experience because it constitutes identity; loss of such control may engender conspiratorial theories.

14. Arthur Koestler, *The Act of Creation* (Harmondsworth, England: Arkana, 1989 [1964]), p. 35. The intersection of the two planes may be a collision (which explodes the accumulated tension and produces laughter), or fusion (intellectual synthesis, creating curiosity), or a prolonged confrontation (the slow catharsis of aesthetic or tragic experience, often yielding tears). If the clash of matrices is limited to humor, the content of which always reflects some form of anxiety, then the power-oriented self-asserting impulse, robbed of its target, is simply released in laughter. With the tragic, aesthetic, or curious, however, there is a prolonged suspension of awareness between the two planes of perception, generating self-transcendence. Perhaps, like the depth of three-dimensional vision or stereophonic sound, which comes from the juxtaposition of two eyes or two ears, three-dimensional consciousness, or reflexivity, arises from the intersection of two fields in the brain, the simplest example being hemispheric specialization. In Jungian psychology, the awareness of opposites is the "specific feature of consciousness." See Edinger, *Ego and Archetype: Individuation and the Religious Function of the Psyche* (Baltimore: Penguin Books, 1973), p. 18.

15. Mihaly Csikszentmihalyi, *Creativity: Flow and the Psychology of Discovery and Invention* (New York: HarperCollins, 1996), pp. 57, 156, 329, 346. Csikszentmihalyi found that such people experience both poles "with equal intensity and without inner conflict," moving from one extreme to the other as the occasion requires. He also noted that their stereoscopic perception was often rooted in a parallactic

childhood, as when two parents have conflicting value systems. An exponential effect would then seem inherent in the cosmopolitan nature of modern society.

16. Paul Ornstein, ed., *The Search for the Self: Selected Writings of Heinz Kohut, 1950–1978*, 2 vols. (New York: International Universities Press, 1978), 1: 271–274.

17. Evidence for the illusion of conscious control is found in hypnotic experiences, in the sincere but bogus explanations that split-brain patients give for their actions, and in experiments such as that conducted by neurophysiologist Benjamin Libet in which subjects who were wired to electrodes and asked to flex a finger registered a flurry of brain activity a fraction of a second *before* the conscious mind sent the "order" to flex. See Timothy Ferris, *The Mind's Sky: Human Intelligence in a Cosmic Context* (New York: Bantam Books, 1992, pp. 74–75).

18. See Guy Claxton, *Hare Brain, Tortoise Mind: Why Intelligence Increases When You Think Less* (Hopewell, N.J.: Ecco Press, 1997); and John S. Dacey and Kathleen H. Lennon, *Understanding Creativity: The Interplay of Biological, Psychological, and Social Factors* (San Francisco: Jossey-Bass, 1998). At the extreme, Claxton suggests, the lost vitality is sought in violence, pornography, or drugs. David Bohm adds that the inner conflicts rooted in an authoritarian childhood can lead to an avoidance of self-reflection. Every time the mind tries to focus on its contradictions, it jumps to something else. "Either it continues to dart from one thing to another, or it reacts with violent excitement that limits all attention to some triviality, or it becomes dead, dull, or anesthetized,

or it projects fantasies that cover up all the contradictions" (*On Creativity* [New York: Routledge, 1998], p. 21); see also Silvano Arieti, *Creativity: The Magic Synthesis* (New York: Basic Books, 1976). The reduction of symbols to signs is treated in depth in Jungian psychology.

19. Ernest Hartmann, *Boundaries in the Mind: A New Psychology of Personality* (New York: Basic Books, 1991). The typical "thick-boundaried" person tends to be conventional, analytical, focused, and ambitious, but inflexible and lacking imagination and intimacy. Frequently older and male, he is more likely to be found in politics, business, law, or engineering. The "thin-boundaried," on the other hand, while creative, intuitive, sensitive, and open, tend to be unfocused, disorganized, and less stable. They are typically young, more androgynous, and heavily represented among artists, teachers, and therapists.

I regret that space allows only crude schematic summaries of these psychological models. There is no substitute for reading the original works. The poles, of course, should not be seen as inferior/superior but rather as complementary, a symbiosis within the human community, each serving an indispensable function.

The correlation of early responsibility, thick boundaries, and extraversion recalls David McClelland's study of the achieving personality, which linked "independence training" with high levels of financial success (*The Achieving Society* [Princeton: D. Van Nostrand, 1961]).

In *Hare Brain, Tortoise Mind* (pp. 151–153), Claxton expands on something very similar to Hartmann's "synaptic freeways" (*Boundaries*, ch. 12). The idea is also reinforced by the suggestion that the superiority of the human mind to that of other primates may correlate with the fact that the appearance of the mature human bears an astonishing resemblance to the eight-month fetus of a chimpanzee. The delay in human maturation, responsible for the helplessness of the infant but also for extended postnatal development, parallels Hartmann's hypothesis that impermanent brain pathways may enhance creativity. Unlike that of the chimpanzee, the human skull is born with the sutures still wide open, allowing continued expansion of the neocortex. In fact, three-quarters of the skull growth takes place after birth. Thus, while the chimp is born with a highly specific instinctual code that integrates him immediately into his forest environment, the human is left to solve many of his environmental challenges with the nonspecific processes of conscious thought. Konrad Lorenz suggests that curiosity is related to this trade-off: "Human exploratory inquisitive behavior—restricted in animals to a brief developmental phase—is extended to persist until the onset of senility" (*Studies in Animal and Human Behavior* [London: Methuen, 1971], p. 239). Robert Bly adds that electronic culture now hinders the normal development of this exploratory function, resulting in a loss of confidence and a feeling of insecurity, which leads to a deliberate numbing of probing capacities (*The Sibling Society* [Reading, Mass.: Addison-Wesley, 1996], p. 59). It is one more explanation for contemporary apathy toward the exploration of space.

20. This discussion is based largely on Robert Ornstein, *The Right Mind: Making Sense of the Hemispheres* (New York: Harcourt Brace, 1997); Richard J. Davidson and Kenneth Hugdahl, eds., *Brain Asymmetry* (Cambridge, Mass.: MIT Press, 1995); D. Frank Benson and Eran Zaidel, eds., *The Dual Brain: Hemispheric Specialization in Humans* (New York: Guilford Press, 1985); Sally P. Springer and Georg Deutsch, *Left Brain, Right Brain,* 4th ed. (New York: W. H. Freeman, 1993); Sid J. Segalowitz, *Two Sides of the Brain: Brain Lateralization Explored* (Englewood Cliffs, N.J.: Prentice-Hall, 1983); and Ann Moir and David Jessel, *Brain Sex: The Real Difference Between Men and Women* (New York: Dell, 1991). The later part of my discussion is of course speculative. There are many possible explanations, such as an individual whose left dominance (characteristic of the average male) is countered by a thicker corpus collosum (characteristic of the average female). It is the balanced polarization that seems essential. On the correlation of interhemispheric communication with creativity, see Dacey and Lennon, *Understanding Creativity*, p. 211.

21. The theme of C. P. Snow's celebrated Rede Lecture deploring the separation between the literary and scientific cultures (published as *The Two Cultures and the Scientific Revolution* [New York: Cambridge University Press, 1959]) has found a resurgence in works like Edward O. Wilson's bestselling *Consilience: The Unity of Knowledge* (New York: Vintage, 1999), Ken Wilber's *The Marriage of Sense and Soul: Integrating Science and Religion* (New York: Random House, 1998), and Paul T. Brockelman's *Cosmology and Creation:*

The Spiritual Significance of Contemporary Cosmology (New York: Oxford University Press, 1999).

22. The media, particularly music and film, have become the sacred canopy for many. In the postmodern void, the overarching context of value, purged of religious consensus, defaults to the marketplace. Thus, while the stars in the sky symbolize the transcendent reality for some, for others it is the "stars" in the media. The desire in both cases is to connect in some way to what is sincerely perceived as the outermost sphere of meaning, be it the night sky or the aura of celebrity. The depth of the symbol corresponds to the level of self-awareness.

23. Richard Tarnas, *The Passion of the Western Mind: Understanding the Ideas That Have Shaped Our World View* (New York: Ballantine Books, 1991), pp. 442–444. Gareth S. Hill, in *Masculine and Feminine: The Natural Flow of Opposites in the Psyche* (Boston: Shambala, 1992), suggests a quaternity in place of Tarnas's duality, the masculine and feminine each having a static and a dynamic pole. The static-feminine/dynamic-masculine polarity corresponds to the more traditional transcendent-maternal/aggressive-hero concept of gender; the tension of Western culture, however, lies more on the static-masculine/dynamic-feminine axis, corresponding to Apollonian/Dionysian, senex/puer, or preservation/renewal.

24. Jacob Needleman, *A Sense of the Cosmos: The Encounter of Modern Science and Ancient Truth* (New York: Arkana, 1988).

25. An excellent synthesis of the new discoveries in nonlinear systems dynamics is Fritjof Capra, *The Web of Life: A*

New Scientific Understanding of Living Systems (New York: Anchor Books, 1996).

26. From "The Last Judgement," quoted in Arthur C. Clarke, ed., *The Coming of the Space Age* (New York: Meredith Press, 1967), p. 301.

27. Mailer, *Of a Fire on the Moon*, pp. 412–413.

28. William Irwin Thompson, *Passages About Earth: An Exploration of the New Planetary Culture* (New York: Harper & Row, 1974), p. 5.

29. Daniel Boorstin, *Cleopatra's Nose: Essays on the Unexpected* (New York: Vintage, 1995), p. 17.

30. Chet Raymo, *The Soul of the Night: An Astronomical Pilgrimage* (Englewood Cliffs, N.J.: Prentice-Hall, 1985), p. ix.

Chapter Five

1. See Robert Jay Lifton and Eric Olson, *Living and Dying* (New York: Bantam, 1975), ch. 3; and Robert Jay Lifton, *The Life of the Self: Toward a New Psychology* (New York: Basic Books, 1983), ch. 2.

2. Paraphrasing a letter from Carl Sagan, Ithaca, N.Y., to Buzz Aldrin, Laguna Beach, Calif., March 1, 1994.

Bibliography

Aldiss, Brian. *Trillion Year Spree: The History of Science Fiction.* New York: Atheneum, 1986.

Aldiss, Brian, and Harry Harrison, eds. *Hell's Cartographers: Some Personal Histories of Science Fiction Writers.* London: Future, 1976.

Aldrin, Buzz, and Malcolm McConnell. *Men from Earth.* New York: Bantam, 1989.

Aldrin, Buzz, and Wayne Warga. *Return to Earth.* New York: Random House, 1973.

American Astronautical Society. AAS History Series. 20 vols. San Diego: American Astronautical Society, 1980–1997.

"Apollo's Moon Mission: Here Are the Results." *U.S. News and World Report*, August 4, 1969, pp. 24–27.

Arendt, Hannah. *The Human Condition.* Chicago: University of Chicago Press, 1958.

————. "Man's Conquest of Space." In *America, Changing . . .* Edited by Patrick Gleeson. Columbus, Ohio: Charles E. Merrill, 1968.

Arieti, Silvano. *Creativity: The Magic Synthesis.* New York: Basic Books, 1976.

Armstrong, Neil, Edwin Aldrin, and Michael Collins. *First on the Moon*. Boston: Little, Brown, 1970.

Associated Press and John Barbour. *Footprints on the Moon*. N.p.: Associated Press, 1969.

Atwill, William D. *Fire and Power: The American Space Program as Postmodern Narrative*. Athens, Ga.: University of Georgia Press, 1994.

Auden, W. H. "Moon Landing." *Selected Poems*. Edited by Edward Mendelson. New York: Vintage, 1989.

Bainbridge, William Sims. *The Spaceflight/Revolution: A Sociological Study*. Malabar, Fla.: Robert E. Krieger, 1983 [1976].

Baker, David. *The History of Manned Space Flight*. New York: Crown Publishers, 1982.

———. *Spaceflight and Rocketry: A Chronology*. New York: Facts on File, 1996.

Barbree, Jay, and Martin Caidin. *Destination Mars in Art, Myth, and Science*. New York: Penguin Putnam, 1997.

Barfield, Owen. *The Rediscovery of Meaning and Other Essays*. Middletown, Conn.: Wesleyan University Press, 1977.

Barrett, William. *Time of Need: Forms of the Imagination in the Twentieth Century*. New York: Harper & Row, 1972.

Barth, Hans. *Hermann Oberth: Leben, Werk und Auswirkung auf die spätere Raumfahrtentwicklung*. N.p.: Uni-Verlag, 1985.

Benson, D. Frank, and Eran Zaidel, eds. *The Dual Brain: Hemispheric Specialization in Humans*. New York: Guilford Press, 1985.

Bleiler, E. F., ed. *Science Fiction Writers: Critical Studies of the Major Authors from the Early Nineteenth Century to the Present Day*. New York: Charles Scribner's Sons, 1982.

Bly, Robert. *The Sibling Society*. Reading, Mass.: Addison-Wesley, 1996.

Bohm, David. *On Creativity*. New York: Routledge, 1998.

Boorstin, Daniel. *Cleopatra's Nose: Essays on the Unexpected*. New York: Vintage, 1995.

Borman, Frank, and Robert J. Serling. *Countdown: An Autobiography*. New York: William Morrow, 1988.

Bova, Ben. *The High Road*. Boston: Houghton Mifflin, 1981.

Bradbury, Ray, Walter Sullivan, Carl Sagan, Bruce Murray, and Arthur C. Clarke. *Mars and the Mind of Man*. New York: Harper & Row, 1973.

Braun, Wernher von, Frederick I. Ordway, III, and Dave Dooling. *Space Travel: A History; An Update of "History of Rocketry and Space Travel."* New York: Harper & Row, 1985.

Burrows, William E. *Exploring Space: Voyages in the Solar System and Beyond*. New York: Random House, 1990.

———. *This New Ocean: The Story of the First Space Age*. New York: Random House, 1998.

Campbell, Joseph. *Myths to Live By*. New York: Bantam, 1973.

Capra, Fritjof. *The Web of Life: A New Scientific Understanding of Living Systems*. New York: Anchor Books, 1996.

Carpenter, M. Scott, et al. *We Seven*. New York: Simon & Schuster, 1962.

Carter, Dale. *The Final Frontier: The Rise and Fall of the American Rocket State*. London: Verso, 1988.

Caspar, Max. *Kepler*. New York: Dover Publications, 1993.

Cernan, Eugene, and Don Davis. *The Last Man on the Moon*. New York: St. Martin's Press, 1999.

Chaikin, Andrew. *A Man on the Moon: The Voyages of the Apollo Astronauts.* New York: Viking Press, 1994.

———. "'Columbia,' Flight Controllers, and the Future." *Sky and Telescope* 61 (June 1981): 485–487.

Clark, Kenneth. *Civilisation: A Personal View.* New York: Harper & Row, 1969.

Clarke, Arthur. C., ed. *The Coming of the Space Age.* New York: Meredith Press, 1967.

———. *Profiles of the Future: An Inquiry into the Limits of the Possible.* New York: Harper & Row, 1973.

———. *The Promise of Space.* New York: Harper & Row, 1968.

———. *The View from Serendip.* New York: Ballantine Books, 1978.

———. *Voices from the Sky: Previews of the Coming Space Age.* New York: Harper & Row, 1965.

Claxton, Guy. *Hare Brain, Tortoise Mind: Why Intelligence Increases When You Think Less.* Hopewell, N.J.: Ecco Press, 1997.

Collins, Martin, and Sylvia K. Kraemer, eds. *Space: Discovery and Exploration.* N.p.: Hugh Lauter Levin Associates, 1993.

Collins, Michael. *Carrying the Fire: An Astronaut's Journeys.* New York: Bantam, 1983 [1974].

———. *Liftoff: The Story of America's Adventure in Space.* New York: Grove Press, 1988.

Columbia Broadcasting System. CBS News. *10:56:20 PM EDT, 7/20/69: The Historic Conquest of the Moon as Reported to the American People by CBS News over the CBS Television Network.* New York: CBS, 1970.

Cooper, Henry S. F. "A Reporter at Large: Men on the Moon (I—Just One Big Rockpile)." *New Yorker*, April 12, 1969, pp. 53–90.

———. *Thirteen: The Flight That Failed*. New York: Dial Press, 1973.

Cronin, Vincent. *The View from Planet Earth*. New York: Morrow, 1981.

Crouch, Tom D. *Aiming for the Starts: The Dreamers and Doers of the Space Age*. Washington, D.C.: Smithsonian Institution Press, 1999.

Crum, William L. *Lunar Lunacy and Other Commentaries*. Philadelphia: Dorrance & Co., 1965.

Csikszentmihalyi, Mihaly. *Creativity: Flow and the Psychology of Discovery and Invention*. New York: HarperCollins, 1996.

Curtis, Anthony R. *Space Almanac: Facts, Figures, Names, Places, Lists, Charts, Tables, Maps Covering Space from Earth to the Edge of the Universe*. Woodsboro, Md.: Arcsoft Publishers, 1990.

Dacey, John S., and Kathleen H. Lennon. *Understanding Creativity: The Interplay of Biological, Psychological, and Social Factors*. San Francisco: Jossey-Bass, 1998.

Davidson, Richard J. and Kenneth Hugdahl, eds. *Brain Asymmetry*. Cambridge, Mass.: MIT Press, 1995.

Dewaard, E. John and Nancy Dewaard. *History of NASA: America's Voyage to the Stars*. New York: Exeter Books, 1984.

Dornberger, Walter. *V 2*. London: Hurst & Blackett, 1954.

Dubb, Edwin. "Without Earth There Is No Heaven: The Cosmos Is Not a Physicist's Equation." *Harper's Magazine* 289 (February 1995): 33–41.

Durant, Frederick C., and Ron Miller. *Worlds Beyond: The Art of Chesley Bonestell*. Norfolk, Va.: Donning, 1983.

Edinger, Edward. *Ego and Archetype: Individuation and Religious Function in the Psyche*. Baltimore: Penguin Books, 1973.

Eisele, Thomas D. "'The *Eagle* Has Landed': Symbol of Savagery, Symbol of Serenity." *Michigan Quarterly Review* 18 (Spring 1979): 177–185.

Eiseley, Loren. *The Invisible Pyramid*. New York: Charles Scribner's Sons, 1970.

———. *The Unexpected Universe*. San Diego: Harvest Books, 1969.

Eliot, T. S. *Four Quartets*. New York: Harcourt Brace Jovanovich, 1971.

Elliot, Jeffrey M., ed. *The Future of the Space Program/Large Corporations and Society: Discussions with 22 Science Fiction Writers*. San Bernardino, Calif.: Borgo Press, 1981.

Etzioni, Amitai. *The Moon-Doggle: Domestic and International Implications of the Space Race*. Garden City, N.Y.: Doubleday, 1964.

Fallaci, Oriana. *If the Sun Dies*. New York: Atheneum, 1967.

Fenton, Robert W. *The Big Swingers*. Englewood Cliffs, N.J.: Prentice-Hall, 1967.

Ferris, Timothy. *The Mind's Sky: Human Intelligence in a Cosmic Context*. New York: Bantam Books, 1992.

Finneran, Richard J., ed. *The Collected Poems of W. B. Yeats*. New York: Collier Books, 1989.

Fitchen, John. *The Construction of Gothic Cathedrals: A Study of Medieval Vault Erection*. Chicago: University of Chicago Press, 1961.

"Flat-Earth Liberals." *National Review*, July 29, 1969, pp. 737–738.

Geis, Larry, and Fabrice Florin, eds. *Moving into Space: The Myths and Realities of Extraterrestrial Space*. New York: Harper & Row, 1980.

Gibson, Edward, ed. *The Greatest Adventure*. Sydney, Aust.: C. Pierson, 1994.

Goldstein, Laurence. "The End of All Our Exploring: The Moon Landing and Modern Poetry." *Michigan Quarterly Review* 18 (Spring 1979): 192–217.

———. *The Flying Machine and Modern Literature*. Bloomington, Ind.: Indiana University Press, 1986.

Gray, Mike. *Angle of Attack: Harrison Storms and the Race to the Moon*. New York: Penguin, 1994.

Grossinger, Richard. *The Night Sky*. San Francisco: Sierra Club Books, 1981.

Hallyn, Fernand. *The Poetic Structure of the World: Copernicus and Kepler*. New York: Zone Books, 1990.

Hardersen, Paul S. *The Case for Space: Who Benefits from Explorations of the Last Frontier?* Shrewsbury, Mass.: ATL Press, 1997.

Harford, James. *Korolev: How One Man Masterminded the Soviet Drive to Beat America to the Moon*. New York: John Wiley & Sons, 1997.

Harman, Willis, and Howard Rheingold. *Higher Creativity: Liberating the Unconscious for Breakthrough Insights*. Los Angeles: Jeremy P. Tarcher, 1984.

Hartmann, Ernest. *Boundaries in the Mind: A New Psychology of Personality*. New York: Basic Books, 1991.

Heinlein, Robert. "Universe." Expanded as *Orphans of the Sky* (New York: Putnam, 1964).

Held, George. "Men on the Moon: American Novelists Explore Lunar Space." *Michigan Quarterly Review* 18 (Spring 1979): 318–342.

Heppenheimer, T. A. *Countdown: A History of Spaceflight.* New York: John Wiley & Sons, 1997.

———. *Toward Distant Suns.* New York: Fawcett Columbine, 1979.

Hill, Gareth S. *Masculine and Feminine: The Natural Flow of Opposites in the Psyche.* Boston: Shambala, 1992.

Hillman, James. *Revisioning Psychology.* New York: Harper & Row, 1975.

Holtsmark, Erling B. *Edgar Rice Burroughs.* Boston: Twayne, 1986.

Hotz, Robert Lee. "Apollo's Unseen Titan." *Los Angeles Times*, July 3, 1994, pp. 1, 28, 31.

Hoyt, William Graves. *Lowell and Mars.* Tucson, Ariz.: University of Arizona Press, 1976.

Hurt, Harry III. *For All Mankind.* New York: Atlantic Monthly Press, 1988.

Hyers, M. Conrad. "Ambivalent Man and His Ambiguous Moon: Apollo 11 in Mythological Perspective." *The Christian Century*, September 10, 1969, pp. 1158–1162.

Irwin, James B. and William A. Emerson, Jr. *To Rule the Night: The Discovery Voyage of Astronaut Jim Irwin.* New York: A. J. Holman, 1973.

Jastrow, Robert. *Journey to the Stars: Space Exploration—Tomorrow and Beyond.* New York: Bantam, 1989.

Keen, Sam. *Apology for Wonder.* New York: Harper & Row, 1969.

Kiernan, Vincent. "Study Finds Space Support Dwindling." *Space News,* February 27–March 5, 1995, p. 6.

Kluger, Jeffrey. *The Apollo Adventure: The Making of the Apollo Space Program and the Movie* Apollo 13. New York: Pocket Books, 1995.

———. *Journey Beyond Selene: Remarkable Expeditions Past Our Moon and to the Ends of the Solar System.* New York: Simon & Shuster, 1999.

Koestler, Arthur. *The Act of Creation.* Harmondsworth, England: Arkana, 1989.

———. *The Watershed: A Biography of Johannes Kepler.* Lanham, Md.: University Press of America, 1960.

Kranz, Eugene. "Lessons Learned: Americans in Space." *Sky and Telescope* 62 (October 1982): 313–316.

Kudlay, Robert R., and Joan Leiby. *Burroughs' Science Fiction.* Geneseo, N.Y.: School of Library and Information Science, State University College of Arts and Science, 1973.

Lasby, Clarence G. *Project Paperclip: German Scientists and the Cold War.* New York: Atheneum, 1971.

Lavery, David. *Late for the Sky: The Mentality of the Space Age.* Carbondale, Ill.: Southern Illinois University Press, 1992.

Lehman, Milton. *Robert H. Goddard: Pioneer of Space Research.* New York: Da Capo Press, 1988.

Lewis, Richard S. *Appointment on the Moon.* New York: Ballantine Books, 1968.

Ley, Willy. *Rockets, Missiles, and Space Travel.* New York: Viking Press, 1957.

Ley, Willy, and Chesley Bonestell. *The Conquest of Space*. New York: Viking Press, 1949.

Lifton, Robert Jay. *The Life of the Self: Toward a New Psychology*. New York: Basic Books, 1983.

Lifton, Robert Jay, and Eric Olson. *Living and Dying*. New York: Bantam, 1975.

Lindbergh, Anne Morrow. *Earth Shine*. New York: Harcourt, Brace & World, 1969.

Lorenz, Konrad. *Studies in Animal and Human Behavior*. London: Methuen, 1971.

Lovell, Jim, and Jeffrey Kluger. *Lost Moon: The Perilous Voyage of Apollo 13*. Boston: Houghton Mifflin, 1994.

Lovelock, J. E. *Gaia: A New Look at Life on Earth*. New York: Oxford University Press, 1982 [1979].

Lupoff, Richard A. *Edgar Rice Burroughs: Master of Adventure*. New York: Canaveral Press, 1965.

McCurdy, Howard E. *Space and the American Imagination*. Washington, D.C.: Smithsonian Institution, 1997.

McDougall, Walter A. *The Heavens and the Earth: A Political History of the Space Age*. New York: Basic Books, 1985.

MacLeish, Kenneth. "Legacy from the Age of Faith: Chartres." *National Geographic* 136 (December 1969): 857–882.

McClelland, David. *The Achieving Society*. Princeton: D. Van Nostrand, 1961.

Mailer, Norman. *Of a Fire on the Moon*. New York: New American Library, 1971.

Maslow, Abraham H. *The Farther Reaches of Human Nature*. New York: Viking Press, 1971.

————. *Religions, Values, and Peak Experiences.* New York: Viking Press, 1970.

————. *Toward a Psychology of Being.* 2nd ed. New York: Van Nostrand, 1968.

Mazlish, Bruce. "The Idea of Progress." *Daedalus* 92 (Summer 1963): 447–461.

Mazlish, Bruce, ed. *The Railroad and the Space Program: An Exploration in Historical Analogy.* Cambridge, Mass.: MIT Press, 1965.

Medawar, P. B. *The Hope of Progress.* London: Methuen, 1972.

Mendelssohn, Kurt. *The Riddle of the Pyramids.* London: Thames & Hudson, 1974.

Miller, Ron. *The Dream Machines: An Illustrated History of the Spaceship in Art, Science and Literature.* Malabar, Fla.: Krieger, 1993.

Mitchell, Edgar, and Dwight Williams. *The Way of the Explorer: An Apollo Astronaut's Journey through the Material and Mystical Worlds.* New York: G. P. Putnam's Sons, 1996.

Mohs, Mayo. "God, Man and Apollo." *Time,* January 1, 1973, pp. 50–51.

Moir, Ann, and David Jessel. *Brain Sex: The Real Difference between Men and Women.* New York: Dell, 1991.

"The Moon Landing Revisited." *The American Enterprise* (July/August 1994), pp. 88–91.

"Moonraker." *Omni* 11 (July 1989): 18–19.

Moser, Leo J. *The Technology Trap: Survival in a Man-made Environment.* Chicago: Nelson-Hall, 1979.

Moskowitz, Sam. *Seekers of Tomorrow: Masters of Modern Science Fiction.* Cleveland: World Publishing Co., 1966.

Mumford, Lewis. *The Myth of the Machine: Technics and Human Development*. New York: Harcourt, Brace & World, 1967.

————. *The Myth of the Machine: The Pentagon of Power*. New York: Harcourt Brace Jovanovich, 1970.

Murray, Bruce. *Journey into Space: The First Three Decades of Space Exploration*. New York: W. W. Norton, 1989.

Murray, Charles, and Catherine Bly Cox. *Apollo: The Race to the Moon*. New York: Simon & Schuster, 1989.

National Aeronautics and Space Administration. *Why Man Explores*. Washington, D.C.: U.S. Government Printing Office, 1976. [Symposium: James Michner, Norman Cousins, Philip Morrison, Jacques Cousteau, Ray Bradbury]

Neal, Valarie, ed. *Where Next, Columbus? The Future of Space Exploration*. New York: Oxford University Press, 1994.

Needell, Allan A., ed. *The First 25 Years in Space: A Symposium*. Washington, D.C.: Smithsonian Institution Press, 1983.

Needleman, Jacob. *A Sense of the Cosmos: The Encounter of Modern Science and Ancient Truth*. New York: Arkana, 1988.

Neufeld, Michael J. *The Rocket and the Reich: Peenemünde and the Coming of the Ballistic Missile Era*. Cambridge, Mass.: Harvard University Press, 1995.

Nicolson, Marjorie Hope. *Voyages to the Moon*. New York: Macmillan, 1960.

Notestein, Wallace. *The English People on the Eve of Colonization, 1603–1630*. New York: Harper & Row, 1954.

Oberg, James, and Alcestis Oberg. *Pioneering Space: Living on the Next Frontier*. New York: McGraw-Hill, 1986.

Oberth, Hermann. *Man into Space: New Projects for Rocket and Space Travel.* London: Weidenfeld & Nicolson, 1957.

Oneill, Gerard. *The High Frontier: Human Colonies in Space.* New York: William Morrow, 1977.

Ordway, Frederick I. III, and Randy Liebermann. *Blueprint for Space: Science Fiction to Science Fact.* Washington, D.C.: Smithsonian Institution Press, 1992.

Ordway, Frederick I. III, and Mitchell R. Sharpe. *The Rocket Team.* Cambridge, Mass.: MIT Press, 1982.

Ornstein, Paul, ed. *The Search for the Self: Selected Writings of Heinz Kohut: 1950–1978.* 2 Vols. New York: International Universities Press, 1978.

Ornstein, Robert. *The Right Mind: Making Sense of the Hemispheres.* New York: Harcourt Brace, 1997.

"Overlooked Space Program Benefits." *Aviation Week.* March 15, 1971, p. 11.

Pauli, Wolfgang. "The Influence of Archetypal Ideas on the Scientific Theories of Kepler." In *The Interpretation of Nature and the Psyche.* Edited by the Bollingen Foundation. New York: Pantheon Books, 1955.

Porges, Irwin. *Edgar Rice Burroughs: The Man Who Created Tarzan.* Provo, Utah: Brigham Young University Press, 1975.

Postman, Neil. *Amusing Ourselves to Death: Public Discourse in the Age of Show Business.* Harmondsworth, England: Penguin, 1986.

Raymo, Chet. *The Soul of the Night: An Astronomical Pilgrimage.* Englewood Cliffs, N.J.: Prentice-Hall, 1985.

Roland, Alex, ed. *A Spacefaring People: Perspectives on Early Spaceflight.* Washington, D.C.: NASA, 1985.

Russell, Peter. *The Global Brain: Speculations on the Evolutionary Leap to Planetary Consciousness.* Los Angeles: J. P. Tarcher, 1983.

Sagan, Carl. Letter to Buzz Aldrin, Laguna Beach, Calif., March 1, 1994. Author's private correspondence.

———. *Broca's Brain: Reflections on the Romance of Science.* New York: Random House, 1979.

———. *Cosmos.* New York: Random House, 1980.

———. *Murmurs of Earth: The Voyager Interstellar Record.* New York: Ballantine Books, 1978.

———. *Pale Blue Dot: A Vision of the Human Future in Space.* New York: Random House, 1994.

Sanford, Charles L. "An American Pilgrim's Progress." *American Quarterly* 4 (Winter 1955): 297–310.

———. *The Quest for Paradise: Europe and the American Moral Imagination.* Urbana, Ill.: University of Illinois Press, 1961.

Schachtel, Ernest. *Metamorphosis: On the Development of Affect, Perception, Attention, and Memory.* New York: Basic Books, 1984.

Schefter, James. *The Race: The Uncensored Story of How America Beat Russia to the Moon.* New York: Doubleday, 1999.

Schick, Ron, and Julia Van Haaften. *The View from Space: American Astronaut Photography, 1962–1972.* New York: Clarkson N. Potter, 1988.

Schmitt, Harrison. "Exploring Taurus-Littrow." *National Geographic* 144 (September 1973): 290–325.

———. "The New Ocean of Space." *Sky and Telescope* 64 (October 1982): 327–329.

Schwartz, S. G. "Amour De Voyage." *Michigan Quarterly Review* 18 (Spring 1979): 266.

Scott, David R. "What Is It Like to Walk on the Moon?" *National Geographic* 144 (September 1973): 326–331.

Segalowitz, Sid J. *Two Sides of the Brain: Brain Lateralization Explored.* Englewood Cliffs, N.J.: Prentice-Hall, 1983.

Shapiro, Kenneth J., and Irving E. Alexander. *The Experience of Introversion.* Durham, N.C.: Duke University Press, 1975.

Shepard, Alan, and Deke Slayton. *Moon Shot: The Inside Story of America's Race to the Moon.* Atlanta: Turner Publishing, 1994.

Simson, Otto von. *The Gothic Cathedral: Origins of Gothic Architecture and the Medieval Concept of Order.* 3rd ed. (Bollingen Series XLVIII). Princeton: Princeton University Press, 1988.

Slayton, Donald K., and Michael Cassutt. *Deke! U.S. Manned Space: From Mercury to the Shuttle.* New York: Tom Doherty Associates, 1994.

Snow, C. P. "The Moon Landing." *Look,* August 26, 1969, pp. 68–70, 72.

"Solar System: It Is Modeled in Miniature by Saturn, Its Rings and Nine Moons." *Life,* May 29, 1944, pp. 78–80, 83–84, 86.

Springer, Sally, and Georg Deutsch. *Left Brain, Right Brain.* 4th ed. New York: W. H. Freeman, 1993.

Stanford, Donald H. "The Moon Landing: A Psychoanalytical Interpretation." *Michigan Quarterly Review* 18 (Spring 1979): 220–229.

Stoiko, Michael. *Soviet Rocketry: Past, Present, and Future.* New York: Holt, Rinehart & Winston, 1970.

Stuhlinger, Ernst, and Frederick I. Ordway III. *Wernher von Braun: Crusader for Space.* Malabar, Fla.: Krieger, 1996.

Tarnas, Richard. *The Passion of the Western Mind: Understanding the Ideas That Have Shaped Our World View.* New York: Ballantine Books, 1991.

"The Talk of the Town." *New Yorker,* July 26, 1969, pp. 25–30.

"The Talk of the Town." *New Yorker,* December 30, 1972, pp. 21–24.

Taylor, Humphrey. "The Harris Poll on Space: Stronger Public Support but not for Spending More," *Space Times* 36 (November–December 1997): 15.

Terzian, Yervant, and Elizabeth Bilson, eds. *Carl Sagan's Universe.* Cambridge: Cambridge University Press, 1997.

Thomas, Davis, ed. *Moon: Man's Greatest Adventure.* New York: Harry N. Abrams, n.d.

Thomas, Shirley. *Men of Space.* 8 vols. Philadelphia: Chilton Book Co., 1961–68.

Thompson, William Irwin. *Passages About Earth: An Exploration of the New Planetary Culture.* New York: Harper & Row, 1974.

Tillyard, E. M. W. *The Elizabethan World Picture.* New York: Vintage Books, n.d.

Tsiolkovsky, K. E. *Selected Works.* Moscow: Mir Publishers, 1968.

Vas Dias, Robert, ed. *Inside Outer Space.* New York: Anchor Books, 1970.

Vajk, J. Peter. *Doomsday Has Been Cancelled.* Culver City, Calif.: Peace Press, 1978.

Von Braun, Wernher, Frederick I. Ordway, III, and Dave Dooling. *Space Travel: A History.* New York: Harper & Row, 1985.

Wachhorst, Wyn. "Carl Sagan, Visionary." *Planetary Report* 17 (May/June 1997), 22.

———. "The Dream of Spaceflight: Nostalgia for a Bygone Future." *The Massachusetts Review* 36 (Spring 1995), 7–32.

———. "Kepler's Children." *The Yale Review* 84 (April 1996), 112–131.

———. "Seeking the Center at the Edge: Perspectives of the Meaning of Man in Space." *The Virginia Quarterly Review* 69 (Winter 1993), 3–23.

Wachhorst, Wyn, and Buzz Aldrin. "A Cosmic Voyage: The Dream of Spaceflight." *The Explorer's Journal* 76 (Spring 1998), 14–19.

Walter, William J. *Space Age.* New York: Random House, 1992.

Walters, Helen B. *Hermann Oberth: Father of Space Travel.* New York: Macmillan, 1962.

Warner, Harry, Jr. *All Our Yesterdays: An Informal History of Science Fiction Fandom in the Forties.* Chicago: Advent, 1969.

Warren, Bill. *Keep Watching the Skies: American Science Fiction Films of the Fifties, Vol. 1, 1950–1957.* Jefferson, N.C.: McFarland, 1982.

Weber, Ronald. "Moon Talk." *Journal of Popular Culture* 9 (Summer 1975): 142–152.

————. *Seeing Earth: Literary Responses to Space Exploration.* Athens, Ohio: Ohio University Press, 1985.

Webster, Bayard. "Flight Chief in Houston: Eugene F. Kranz." *New York Times,* March 31, 1982, p. 24.

West, John Anthony. *Serpent in the Sky: The High Wisdom of Ancient Egypt.* New York: Julian Press, 1987.

White, Frank. *The Overview Effect: Space Exploration and Human Evolution.* Boston: Houghton Mifflin, 1987.

Wiebe, Robert H. *The Search for Order, 1877–1920.* New York: Hill & Wang, 1967.

Wilford, John Noble. *We Reach the Moon.* New York: Bantam, 1969.

Winter, Frank H. *Rockets into Space.* Cambridge, Mass.: Harvard University Press, 1990.

————. *Prelude to the Space Age: The Rocket Societies: 1924–1940.* Washington, D.C.: Smithsonian Institution Press, 1983.

Wolfe, Tom. *The Right Stuff.* New York: Farrar, Straus & Giroux, 1979.

Wulforst, Harry. *The Rocketmakers.* New York: Orion Books, 1990.

Zimmerman, Robert. *Genesis: The Story of Apollo 8: The First Manned Flight to Another World.* New York: Four Walls Eight Windows, 1998.

Zubrin, Robert. *The Case for Mars: The Plan to Settle the Red Planet and Why We Must.* New York: The Free Press, 1996.

————. *Entering Space: Creating a Spacefaring Civilization.* New York: Jeremy P. Tarcher, 1999.

Index

Abstract vs. direct
    experience, 58
Adolescence, 16, 18, 19, 38,
    46, 60–61, 85, 89, 95,
    96, 97, 147
A4 rockets, 32
Ahab, 107
Aldrin, Buzz, 92, 176
Alexander, Irving, 140
All-Story Magazine, 22
America, 18, 19
Amundsen, Roald, 68, 81
Anders, Dave, 176
Apollo 11 flight, 64, 68, 85,
    89, 121
    liftoff for, 86–87
Apollo program, 114, 157,
    160
    cost of, 89, 90, 130, 135
    Apollo 1 module fire,
      40–41, 156

Apollo 5 flight, 121
Apollo 8 flight, 81,
    164–165, 176
Apollo 13 flight, 119,
    121–126, 127–128
Apollo 15 flight, 177
Apollo 16 flight, 91, 177
Apollo 17 flight, 88, 116,
    177
Apollo-Saturn rocket, 28,
    83, 85–86
Pad 34, 156, 157
public's
    perception/reaction,
    90–92, 130, 133, 134.
    See also Spaceflight,
    public attitude toward
scrapping of last three
    flights, 130–131
See also Apollo 11 flight
Apollo 13 (film), 128, 129

Archimedean point, 76, 80, 150, 168
Arendt, Hannah, 76, 80
Aristotle, 15, 57
Armstrong, Neil, 94, 106, 121, 130, 176
Art, 154, 160
Asimov, Isaac, 53
Astounding Science Fiction, 44, 50, 174
Astronaut as symbol, 95–96, 111
Astronomy, 69, 101, 179(n1)
Auden, W. H., 166
Autocentric experience, 188(n12)

Bacon, Francis, 9
Balance, 146, 193(n20)
Balboa, Vasco Nuñez de, 11, 12
Berger, Peter, 136
Black holes, 113
Blake, William, 36, 141
Blue Planet, 64
Bly, Robert, 192(n19)
Bohm, David, 190(n18)

Bonestell, Chesley, 44, 48–51, 54, 56–57, 58, 59, 61, 62–63, 65, 66, 67, 81
Boorstin, Daniel, 155
Borman, Frank, 176
Boston Post, 24
Bradbury, Ray, 23
Brahe, Tycho, 5
Brain lateralization, 145–146, 189(n14)
Bridges at Toko-Ri, The, 126
Burroughs, Edgar Rice, 19, 22–24

Calvinism, 95
Campbell, John W., 19, 44, 174
Campbell, Joseph, 80, 92, 115, 153
Cartesian-Newtonian worldview, 72
Cathedrals, 10, 98, 99, 100, 102–105, 153–154, 160
Celestial mechanics, 4
Center/edge, 80, 81, 111, 112, 114–116, 117, 133, 155
Cernan, Gene, 58, 110

Chafee, Roger, 40–41
Change, 8, 18–19, 154, 161, 188(n12)
Chartres Cathedral, 102–105
Childhood, 144, 170, 190(n18)
Chimpanzees, 192(n19)
Christianity, 72. *See also* Cathedrals; God
Cities, 107
Clarke, Arthur, 23, 31, 44, 74
Claxton, Guy, 143, 190(n18)
Cognitive science, 142
Collier's, 67
Collins, Mike, 68, 91
Columbia shuttle flights, 177
Columbus, Christopher, 5, 38, 54, 57, 76, 115
Comets, 2, 19, 170
Complexity, 149, 150, 187(n8)
Computers, 121
Conquest of Space (Ley and Bonestell), 50–51, 62, 67

Contact (film), 169
Control, 142, 143, 144, 188(n13), 190(n17)
Copernicus, Nicolaus, 9, 13, 179(n1)
Cosmic voyages, 8, 18, 21. *See also* Science fiction
Cosmos series, 108–109, 177
Creativity, 62, 141–142, 143, 146, 151, 162, 189(n15), 192(n19)
Cronkite, Walter, 151
Csikszentmihalyi, Mihaly, 189(n15)
Curiosity, 143, 151, 161, 189(n14), 192(n19)

Dark matter, 113
Death, 53, 160, 161, 162
Descartes, René, 9, 72
Despair, 95, 168
Destination Moon (film), 35, 62–67, 74, 182(n2)
Die Rakete zu den Planetenraumen (Oberth), 30
Donati's comet (1858), 19–20
Dornberger, Walter, 32, 33

Dubb, Edwin, 137
Duke, Charles, 110, 177

Ecology movement, 93
Ecstasy, 162
Edinger, Edward, 182(n17)
Education, 134, 161
Eiseley, Loren, 78–79, 104,
    107, 108, 113
Eliot, T. S., 96
Empiricism, 3, 4, 106
Enlightenment era, 16, 17,
    18
Escapism, 19
Europe, 152
Evolution, 150, 151, 155,
    164
Expanding universe, 21
Exploration, 12, 78, 96,
    139, 143, 145, 146,
    147, 150, 151, 152,
    155, 168–169,
    188(n12)
Extraterrestrial life, 7, 21,
    169

Failure of nerve, 9, 128, 168
Fantasy and Science fiction,
    44

Fear, 139, 155
Fire, 153, 164
Flight director's role, 120
For All Mankind
    (documentary), 92
Freedom, 18, 19, 154
French Revolution, 17
From the Earth to the Moon
    (Verne), 17
Frontiers, 18, 152–153
Fuel, 121
Fundamentalism, 72, 73

Gagarin, Yuri, 175
Gaia, 93, 107, 149
Galaxies, 21, 71, 107–108,
    112, 155, 170
Galilei, Galileo, 6, 7, 9,
    13
Gemini missions, 121, 127,
    175
German Rocket Society, 30,
    33
Glenn, John, 175
God, 3, 4, 72, 103, 113, 136
Goddard, Esther, 28, 69
Goddard, Robert, 21, 24–28,
    34, 44, 69
Goddard Space Center, 31

Gothic cathedrals, 102–105.
    See also Cathedrals
Gravity, 4, 6
Great Comet of 1577, 2
Great Mother, 52–53, 54,
    79, 114
Great Pyramid at Giza,
    100–102. *See also*
    Pyramids
Grissom, Gus, 40–41

Haise, Fred, 122
Haldane, J.B.S., 151
Harmony of the World
    (Kepler), 15
Hartmann, Ernest,
    191(n19)
Heaven, 99, 103, 105
Heinlein, Robert, 16, 35, 44,
    63, 65, 66, 67
Hill, Gareth S., 194(n23)
Hillman, James, 182(n17)
Hitler, Adolf, 32
Hubble, Edwin, 21
Human condition, 62, 76,
    78, 89, 148
Human experience, zones
    of, 70–71
Humor, 189(n14)

Iapetus, 47
Idealism, 109, 156
Idealized past, 17, 18, 46
Imagination, 59, 133, 139,
    141, 144, 167
Immortality, 107, 160,
    161–162
Income, 134
Individuals/individualism,
    9, 13, 16, 18, 37, 39,
    111, 147, 161
Intelligent life, quest for,
    169
Internal models, 77, 78
Introversion/extraversion,
    140–141, 188(n13)

Jeffers, Robinson, 153
Journalism, 132
Jupiter, 177

Kalahari Bushmen, 76–77
Kant, Immanuel, 3
Kasputin Yar, 35
Kennedy, Jack, 88, 89
Kepler, Johannes, 1–8, 9,
    10, 12, 13–16, 179(n1)
    death of, 16
    family of, 14–15

laws of planetary
motion, 3–4, 5–6
Kerwin, Joe, 125
Khrushchev, Nikita, 88
King, Jack, 86
King Kong, 65
Knowledge, 154
Koestler, Arthur, 2, 3, 6, 13,
141
Kohut, Heinz, 143
Korolev, Sergei, 31
Kranz, Gene, 119–121,
123–129, 157

Leonov, Alexei, 36, 37, 38,
39
Ley, Willy, 33, 44, 50
Liberalism, 135
Libet, Benjamin, 190(n17)
Life magazine, 46–47, 49,
50
Lifton, Robert, 161
Lindbergh, Charles, 25
Lippert, Robert, 182(n2)
Lorenz, Konrad, 192(n19)
Los Angeles earthquake
(1994), 133
Love, 117, 164, 171
Lovell, Jim, 122, 176

Lowell, Percival, 19–21, 44,
69
Luna flights, 175, 176

McDougall, Walter, 131
MacLeish, Archibald, 92,
93, 96
Magellan, Ferdinand,
10–11
Mailer, Norman, 88, 152
Mariner flights, 20, 175, 177
Mars, 5, 22–23, 25, 35, 50,
51, 57, 71, 73, 76, 80,
97, 98, 110, 130, 133,
166, 177, 178
canals on, 16, 20–21, 69
*Martian Chronicles
(Bradbury),* 23
Masculinity, 39, 147, 148,
194(n23)
Maslow, Abraham, 138,
139–140
Materialism, 9, 10
Meaning, 10, 12, 16, 37,
45, 53, 57, 75, 89, 92,
94, 110, 114, 136, 137,
140–141, 143, 145,
146, 149, 153, 160,
162, 187(n8), 194(n22)

Means/ends, 89, 110, 132,
139, 154, 161,
188(n12)
Media, 90, 122, 168,
194(n22). *See also*
Television
Medieval period, 2, 9, 10,
15, 19
Mercury-Redstone launch,
121
Merrick, John, 109
Merritt Island, 83
Mimas, 48
Modernity, 2, 12, 17, 19,
39, 73, 152
Moon, 29, 35, 50, 56, 57,
63, 64–65, 68, 69–70,
73–74, 83, 93–94,
96–98, 99, 106, 108,
110, 114, 124, 129,
130, 133, 151, 157,
175, 180(nn 3, 9)
moon shots as hoax, 133
See also Apollo 11 flight;
Apollo program
Morrison, Philip, 76
Mumford, Lewis, 102
Music of the spheres, 4,
180(n3)

Mysterium tremendum,
136–137, 160
Mythology/religion, 93,
115, 153

National Review, 135
Needleman, Jacob, 148
Newton, Isaac, 4, 6, 13, 15,
36, 72
New World, 9
New York Times, 91
"Nightfall" (Asimov), 53
Nixon, Richard, 131, 135

Oberth, Hermann, 17, 25,
28–31, 69
Oort cloud, 170
Optics, 7
Orbits, 5

Paintings, 47–51. *See also*
Bonestell, Chesley
Pal, George, 44, 63, 65,
66–67
Pathfinder, 166, 178
Patriarchy, 147
Pearson, Drew, 135
Peenemünde, 29, 30, 31–32
Phoebe, 47

Pichel, Irving, 63, 65
Play, 110, 139, 152, 154, 155
Plutarch, 180(n3)
Pluto, 21
Polls, 134, 187(n9)
Postman, Neil, 133
Power issues, 134, 138, 139, 154, 161, 189(n14)
Project Moonbase (film), 67
Protestant Ethic, 154
Puritans, 116
Pyramids, 84, 98, 99, 100–102, 106, 109, 153–154, 160

Rational consciousness, 142, 143
Rationalism, 16–17, 147
Renaissance, 4, 152, 167
Ride, Sally, 178
Rockets, 25, 26, 29, 68, 69, 73, 114, 129, 131, 160, 180(n9). *See also individual rocket types*
Rocketship X-M (film), 182(n2)
Romanesque cathedrals, 104

Romanticism, 17, 50, 59, 69, 97, 109

Sagan, Carl, 21, 23, 84, 108–109, 167–170, 177, 178
Sanford, Charles, 116
Satellites, 45
Saturday Evening Post, 66
Saturn, 47–48, 51, 56
Saturn rockets, 28, 85, 87, 153–154
Schmitt, Jack, 95
Science, 4, 72, 73, 76, 88, 93, 106, 133, 146, 149, 150, 154, 169
of Egyptians, 102
scientific revolution, 5, 14
Science fiction, 7, 17, 18, 19, 23, 44, 45–46, 56, 61, 97
fans of, 96, 98
SF films, 63, 65, 74, 85, 105. *See also individual films*
Seashores, 51, 52, 54–55, 56, 62, 78, 84

Self-assertion, 137–138, 141, 147, 149, 189(n14)

Self-awareness, 62, 94, 137, 141, 143, 146, 150, 160, 171, 194(n22)

Seventeenth century, 8–9, 13, 17, 19

SF. *See* Science fiction, SF films

Shapiro, Joel, 140

Shelley, Mary, 17

Shepard, Al, 110, 175

Simulations, 68

Sisyphus, 80

Smith, Charlie, 116

Snow, C. P., 193(n21)

Solar system, 71

Solipsism, 152

Somnium (Kepler), 7–8

Space colonization, 75

Spaceflight, 10, 30, 31, 35, 39, 44–45, 50, 60, 66, 69, 73, 76, 96, 112, 113, 130, 147, 164, 167, 187(n8)
  chronology, 173–178
  public attitude toward, 134, 187(n9), 192(n19). *See also* Apollo

program, public's perception/reaction spending for, 130, 134, 187(n9)

Space race, 31

Spaceship Earth, 84, 107

Space stations, 176, 177, 178

Space walks, 36, 37, 40, 41, 175, 178

Spirituality, 138, 146

Spirit vs. matter, 10, 15, 39, 117, 149, 155

Sputnik, 31, 68, 131

Stanford University, 128

"Star Thrower, The" (Eiseley), 78–79

Stop-motion photography, 66

Stress, 143, 144, 149, 150

Swigert, Jack, 122, 125

Symbols, 72, 75, 95, 99, 105, 123, 136, 143, 147, 160, 161, 167, 194(n22)

Tarnas, Richard, 147–148, 194(n23)

Technology, 106

Telemetry, 28, 121

Telescopes, 7

Television, 131–133, 138.
     See also Media

Tennyson, Alfred, 115

Tereshkova, Valentina, 175

The Thing (film), 65

Things To Come (film), 63

Thirty Years War, 14

Thompson, William, 88,
     154

Thoreau, Henry David,
     90

Time Machine, The (film), 67

Time magazine, 166

Titan, 47, 51, 56

Titanic, 123

Tool-making, 75

Transcendence, 72, 97, 104,
     105, 136, 146, 162,
     194(n22)
   self-transcendence, 137,
     139, 140, 144, 147,
     149, 150, 151, 152,
     189(n14)

Tsiolkovsky, Konstantin, 25,
     69

2001: A Space Odyssey,
     74–75

Ulysses, 115–116

Under the Moons of Mars
     (Burroughs), 22

Van Ronkel, Rip, 63

Venus, 69, 175, 176

Verne, Jules, 17, 19, 29, 44,
     69, 180(n9)

View of earth, 36, 37, 38,
     92–93, 95, 117, 137

Viking flights, 20, 177

von Braun, Wernher, 10,
     30, 32–33, 67, 157

Voyager flights 167,
     169–171, 177

V–2 rockets, 27, 29, 30, 32,
     33–34, 35
   casualties at
     Peenemünde/London,
     34

War of the Worlds, The
     (Wells), 21–22, 24
   as radio broadcast, 69
   as film, 67

Washington Post, 91

Welles, Orson, 69

Wells, H. G., 17, 19, 22, 44,
     69

When Worlds Collide (film), 67

White, Ed, 40–41, 175

White, Frank, 78

White Sands, New Mexico, 27, 34, 35

"Why" questions, 57, 70, 75, 136

Winthrop, John, 116

Woman in the Moon (film), 29, 64–65

Wonder, 39, 73, 110, 111–112, 117, 132, 136–137, 139, 143, 145, 146, 147, 151, 152, 160, 161, 164, 167, 171

World War II, 126

Young, John, 110, 177